Dimensional Analysis

Dimensional Analysis

Practical Guides in Chemical Engineering

Jonathan Worstell

AMSTERDAM • BOSTON • HEIDELBERG • LONDON
NEW YORK • OXFORD • PARIS • SAN DIEGO
SAN FRANCISCO • SINGAPORE • SYDNEY • TOKYO

Butterworth-Heinemann is an imprint of Elsevier

Butterworth-Heinemann is an imprint of Elsevier
225 Wyman Street, Waltham, MA 02451, USA
The Boulevard, Langford Lane, Kidlington, Oxford OX5 1GB, UK

Notice
No responsibility is assumed by the publisher for any injury and/or damage to persons or
property as a matter of products liability, negligence or otherwise, or from any use or
operation of any methods, products, instructions or ideas contained in the material herein.
Because of rapid advances in the medical sciences, in particular, independent verification of
diagnoses and drug dosages should be made.

Library of Congress Cataloging-in-Publication Data
A catalogue record for this book is available from the Library of Congress.

British Library Cataloguing-in-Publication Data
A catalog record for this book is available from the British Library.

ISBN: 978-0-12-801236-9

For information on all Butterworth-Heinemann publications
visit our website at store.elsevier.com

This book has been manufactured using Print On Demand technology. Each copy is produced to
order and is limited to black ink. The online version of this book will show color figures where
appropriate.

Working together
to grow libraries in
developing countries

ELSEVIER   Book Aid
            International

www.elsevier.com • www.bookaid.org

DEDICATION

To Sinta Worstell
Wife and best friend of 40 years

CONTENTS

CHAPTER 1

Introduction

1.1 PROCESS DEVELOPMENT

We generally classify chemical processes by their size: laboratory, pilot plant, or commercial. Process development entails moving technology from the laboratory to the commercial plant. Process development usually begins with an idea that is later successful in laboratory experiments. The proven idea then moves into a pilot plant and if successful there, progresses to a purpose-built commercial plant or into an existing commercial plant. Process support involves moving from an existing commercial plant into a laboratory or pilot plant to solve a problem. We may solve that problem by a small change to the process or with a major overhaul of the process, but, in either case, the solution is first tested in a laboratory or pilot plant, and then confirmed in a test run at the commercial plant.

Process development and process support involve identifying the variables pertinent to and controlling the chemical process, then designing experiments to establish the functional relationship between the variable of interest; i.e., the dependent variable—the variable that will make us money or is costing us money—and the independent variables of the process. These experimental programs must be doable and within the financial ability of the organization sponsoring them.

A good portion of such an experimental program involves collecting information that allows us to move the process from its current size to the next larger size. We call moving from one size to another "scaling." Thus, process development involves upscaling and process support utilizes downscaling and upscaling. Upscaling involves starting with an idea, then proceeding from laboratory to commercial plant. Downscaling involves starting at a commercial plant and conducting laboratory experiments or operating a pilot plant to mimic the commercial plant problem, then upscaling the solution into the commercial plant.

At some point in your chemical engineering career, you will be asked either to develop a process or to provide technical support to an existing process. You are at that point in your career; otherwise, you would not be reading this book.

There are three methods for developing and upscaling a chemical process:

1. Build successively larger plants until you reach a defined commercial size.
2. Derive and solve the various conservation and transport equations describing your process, then use them to size the commercial process.
3. Establish the empirical relationships between the variables of your process, then use similarity to size the commercial process.

During the first 65–75 years of chemical engineering, we employed the first method when developing and upscaling a chemical process. There are several reasons why we used this method. First, during those years, most chemical processes were new and novel and chemical engineers possessed little information about them, particularly with regard to their safety. Therefore, chemical engineers incrementally increased, stepwise, the size of the process and established the operating conditions and monitored the interaction of the chemicals at each completed step. Second, chemical engineers used each successively larger processing unit as an analog computer, sampling the contents of each unit, then plotting the resulting data to obtain the solution to the differential equations describing the chemical process. Third, during those years, chemical and metallurgical engineers had to develop new materials of construction to meet the specifications of their new chemical processes, particularly with regard to high pressure, high temperature, and corrosion resistance. Fourth, physical property databases were rudimentary during those years. In many cases, intermediate-sized processing units; i.e., pilot plants, were built primarily to measure the physical properties of the chemical components comprising the new process. This process development method, while having advantages, is capital intensive, time consuming, and operationally expensive.

The second method arose with the advent of digital computing during the mid-1940s. Digital computing extends the hope that we can step directly from laboratory-sized equipment to commercial-sized equipment via calculation. The realization of this hope requires extensive, robust databases. Thus, the effort during the third and fourth quarters of the twentieth century to establish the physical properties of a wide variety of chemicals and to develop methods for estimating, via calculation, the physical properties of all chemicals. Also, during these years, computing power doubled many times and software became evermore user-friendly. But, even with these advances, the conservation and transport equations for a given chemical process remain difficult to solve numerically. Much of this difficulty arises from the "stiff" differential equations describing the chemical process. Stiff means some of the differential equations have characteristic times much smaller than the other differential equations.[1] Numerical solutions are also difficult to obtain for catalyzed chemical processes. In a catalyzed chemical process, the catalyst is present at parts per million levels while the reactants are present at moles per liter levels. The same constraint occurs when impurities or by-products are included in a digital model. Again, impurities and by-products are present at parts per million levels while reactants and products are present at moles per liter levels. Such sets of differential equations demonstrate the same characteristics of stiff sets of differential equations; namely, numerical calculations will not "close," will not approach a stable result. Today, it is possible to design portions of a chemical process directly from laboratory data, distillation showing the most success, but we are far from designing a complete commercial-sized chemical process via calculation.

Downsizing occurs when a commercial plant exists but a pilot plant for the chemical process does not exist. This situation occurs quite frequently for well-established commercial products, particularly commodity products. Unfortunately, processing issues still arise in such commercial plants, issues which require study in a nonexistent pilot plant. In such cases, you will be asked to design a pilot plant that mimics the commercial plant in order to develop solutions to commercial plant problems. Downsizing a chemical process sounds easy, until you try it. Traditionally, we have simply built a miniature "look-alike" plant when downsizing a commercial plant. This approach depends upon "luck" to reproduce the problem requiring solution. If we alter the controlling regime of the process due to downsizing, then we spend time, capital, and incur operating expense solving a problem related to the pilot plant rather than solving the commercial plant problem. This situation occurs many more times than we care to admit. To successfully downsize a chemical process, we must first identify the regime, either momentum transfer, energy transfer, or mass transfer, causing the problem in the commercial plant, then design the downsized unit to exactly mimic that regime. Otherwise, we expend considerable effort solving an unrelated problem. Due to misidentifying the controlling regime, accurately downsizing a commercial chemical process is more difficult than upscaling it. Thus, downsizing is just as difficult as upscaling a chemical process, if not more so. Chemical engineers have applied digital computing to downsizing also; the success rate is about equivalent to that for upscaling a chemical process.

But, what can we do to successfully develop and support, and subsequently scale, a chemical process if its conservation laws cannot be solved analytically and if the differential equations comprising those laws are "stiff"; i.e., cannot be solved numerically? We could revert to our historic procedure for developing a chemical process; namely, build successively larger processing units until we reach the commercial-sized plant. Today's global economy, however, puts severe constraints upon this particular method of process development and process support.

1.2 TWENTIETH-CENTURY REVOLUTIONS

Three twentieth-century revolutions provide the foundation of our global economy. The transportation revolution includes the invention of the box container, the development of large transport ships with evermore efficient power plants, and the development of highly efficient off-loading facilities. Shipping costs have plummeted during the last 30–40 years because of these technologies. Result: industrial product can be shipped globally at low cost, thereby increasing commercial competition.

The information revolution involves the advent of electronic communications and the Internet. Both have revolutionized commercial marketing, thereby providing every company access to every market ... globally. Thus, every company faces increased competition. With so much competition, the company introducing a process or product to the market makes the most money. In other words, the time from invention to commercialization must be short; otherwise, someone else will beat you to the marketplace and realize your potential profit. Time has become money.

Lastly: the financial revolution, which is based on electronic communications and the Internet. Today, financial markets never close ... they have become one that simply follows the sun as the earth as it rotates on its axis. Thus, there is severe competition for your investment dollars. This competition limits the capital available for investment in process development and process support; i.e., for building pilot plants.

In summary, low transportation costs means product will be manufactured in low-wage countries; thus, the drive to reduce operating costs at all commercial levels and process development expenditures. Competition for capital, in the form of return on investment, means future projects must have returns equal to or greater than those available in the financial market. Thus, the drive to control capital investment in pilot plants during process development. And, lastly, the advent of the Internet means the time from invention to commercialization; i.e., "cycle time," must be reduced. Hence, the drive to minimize the time required to develop a process.

1.3 DIMENSIONAL ANALYSIS

So, how can we develop a chemical process in light of these constraints? Fortunately, there is a mathematical procedure available for upscaling and downscaling chemical processes that does not involve analytically or numerically solving the relevant conservation laws. That mathematical procedure is Dimensional Analysis.

When we upscale or downscale a chemical process, we use conservation laws that describe the changes imposed upon it. Such equations are written descriptively as

$$\text{accumulation} = \text{input} - \text{output} + \text{generation}$$

for a given chemical species. Each term in this descriptive equation contains physical quantities, such as constants, parameters, and variables. Constants fluctuate or vary the least, as their name implies. Constants do not change or vary within the space and during the time we use them. Such constants are inherent in the functioning of the universe. Some common constants are the speed of light in a vacuum, the gravitational constant, Boltzmann's constant, and Planck's constant. Defined constants, such as the permeability of vacuum, are exact. The majority of constants, however, are measured. The speed of light in vacuum and the gravitational constant are examples of measured constants.

Parameters are physical quantities that are unchanging in the context of a physical or chemical process. Parameters are not inherent to the functioning of our universe. They are inherent to the functioning of a given physical or chemical process. Thus, parameters change from one physical or chemical process to another.

The physical quantities that vary the most in a process are, as their name implies, variables. Variables come in two flavors: independent and dependent. Independent variables specify and define physical or chemical processes. Independent variables identify distinguishable requirements, those pertinent interactions between us and the physical or chemical process.[2] Thus, we can change independent variables directly through our interaction with the physical or chemical process. Dependent variables respond to our interaction with the process, to our changing an independent variable.

Dimensional Analysis provides relationships between the dependent variable and a minimum number of independent variables. As chemical engineers, we are introduced to Dimensional Analysis via Lord Rayleigh's Method of Indices and nondimensionalization of differential equations.

Dimensional homogeneity forms the foundation of Lord Rayleigh's Method of Indices. Dimensional homogeneity stipulates that the dimensions on either side of an equality sign must be the same.[3] Lord Rayleigh's Method of Indices is best understood through example.[4] Consider isothermal flow of a viscous Newtonian fluid: what variables influence it and how are they related? From centuries of study, we know that isothermal fluid flow in a pipe depends upon fluid velocity, pipe diameter, fluid density, fluid viscosity, fluid surface tension, if a free surface exists in the pipe during flow, and the force impinging the fluid.

Our first task when undertaking a Dimensional Analysis is to designate the fundamental dimensions we plan to use.[5] For this problem, we will designate length (L), mass (M), and time (T) as our fundamental dimensions. Note that these fundamental dimensions are independent of each other; thus, they can be multiplied or divided by each other. In other words, they can yield "derived" dimensions, such as velocity (L/T) or force (ML/T^2) and so on.

The above variables in terms of fundamental dimensions and derived dimensions are: velocity, v [L/T]; linear dimension, L [L]; force, F [ML/T^2]; density, ρ [M/L^3]; viscosity, μ [M/LT]; surface tension, σ [M/T^2]; and, acceleration of gravity, g [L/T^2]. The bracketed terms identify the fundamental dimensions describing each variable. For this example, we assume the mass in force and the mass in density, viscosity, and surface tension are measured in the same manner. In other words, we will not encumber ourselves with a force mass (M_F) and a weight mass (M_W).[6] Our long history of studying isothermal pipe flow indicates that the flow equation can be expressed as a function such as

$$f(v, L, F, \rho, \mu, \sigma, g) = 0$$

where $f(\ )$ designates an unknown function. If we identify force F as the dependent variable, then the above function becomes

$$\kappa f(v, L, \rho, \mu, \sigma, g) = F$$

where κ is a constant to be determined experimentally.

From dimensional homogeneity, the above function can be expanded as a power series; thus

$$F = \kappa(v)^a (L)^b (\rho)^c (\mu)^d (\sigma)^e (g)^h$$

where a, b, c, ..., h represent unknown powers or indices. In terms of fundamental and derived dimensions, the above power law becomes

$$[MLT^{-2}] = \kappa [LT^{-1}]^a [L]^b [ML^{-3}]^c [L^{-1}MT^{-1}]^d [M\ T^{-2}]^e [LT^{-2}]^h$$

Equating the powers or indices for each fundamental dimension yields

$$M{:}1 = c + d + e$$
$$L{:}1 = a + b - 3c - d + h$$
$$T{:} -2 = -a - d - 2e - 2g$$

Thus, we have six unknowns and three equations. Therefore, we can solve for three unknowns in terms of the remaining three unknowns. Solving for a, b, and c, we obtain

$$a = 2 - d - 2e - 2g$$
$$b = 2 - d - e + g$$
$$c = 1 - d - e$$

Substituting the above into the power law expression for F yields

$$F = \kappa(v)^{(2-d-2e-2g)} (L)^{(2-d-e+g)} (\rho)^{(1-d-e)} (\mu)^d (\sigma)^e (g)^h$$

Grouping similar powers or indices gives

$$F = \kappa(v^2 L^2 \rho) f \left\{ \left(\frac{vL\rho}{\mu} \right)^d, \left(\frac{v^2 L\rho}{\sigma} \right)^e, \left(\frac{v^2}{Lg} \right)^h \right\}$$

Rearranging the above equation yields

$$\frac{F}{v^2 L^2 \rho} = \kappa f \left\{ \left(\frac{vL\rho}{\mu}\right)^d, \left(\frac{v^2 L\rho}{\sigma}\right)^e, \left(\frac{v^2}{Lg}\right)^h \right\}$$

Note that each term in the above equation is dimensionless. The independent dimensionless terms are

$$\left(\frac{vL\rho}{\mu}\right)^d, \left(\frac{v^2 L\rho}{\sigma}\right)^e, \left(\frac{v^2}{Lg}\right)^h$$

We traditionally call such terms dimensionless parameters.[7] The dependent dimensionless term is

$$\frac{F}{v^2 L^2 \rho}$$

We traditionally call such terms dimensionless coefficients.[7]

Note that the indices d, e, and h on the dimensionless parameters are undefined, as is the constant κ. We can assume values for d, e, and h so long as our assumptions ensure that the dimensionless parameters are independent of each other. In the past, engineers have assumed the dimensionless parameters are independent of each other if each parameter contains a unique variable. In the above equation, μ, σ, and g provide a unique variable in each dimensionless parameter to the right of the equality sign. Thus, these dimensionless parameters are independent of each other, which allows us to assume values for d, e, and h. Let us assume

$$d = e = h = 1$$

The function above then becomes

$$\frac{F}{v^2 L^2 \rho} = \kappa f \left\{ \left(\frac{vL\rho}{\mu}\right), \left(\frac{v^2 L\rho}{\sigma}\right), \left(\frac{v^2}{Lg}\right) \right\}$$

which represents the solution to our problem in functional notation. We must perform experiments to determine the functional relationship between the dimensionless coefficient and the dimensionless parameters.

If we know one variable is linearly dependent upon another variable, we need to perform only three experiments to determine the slope and intercept of the function. In general, however, we do not know whether a given variable depends linearly on another variable or not. In such cases, we must perform at least five experiments to determine the curvature of the function or to determine whether the function is curved or oscillatory.[8] Thus, for multivariable functions, the number of experiments to define the function grows rapidly because we must perform five experiments per independent variable while maintaining all other variables constant. Therefore, the total number of experiments $N_{\text{Expts}}^{\text{Total}}$ required to define a function is

$$N_{\text{Expts}}^{\text{Total}} = N_{\text{Expts}}^{M_{\text{IndepVar}}}$$

where N_{Expts} is the number of experiments to be performed per independent variable and M_{IndepVar} is the number of independent variables in the function.

For this example, the dimensionless function is

$$\frac{F}{v^2 L^2 \rho} = \kappa f \left\{ \left(\frac{vL\rho}{\mu}\right)^d, \left(\frac{v^2 L\rho}{\sigma}\right)^e, \left(\frac{v^2}{Lg}\right)^h \right\}$$

Thus, we have one dimensionless coefficient and three dimensionless parameters. The total number of experiments required to define this function is

$$N_{\text{Expts}}^{\text{Total}} = N_{\text{Expts}}^{M_{\text{IndepVar}}} = 5^3 = 125$$

If we solve this example dimensionally, the solution function is

$$F = \kappa * f(v, L, \rho, \mu, \sigma, g)$$

and $N_{\text{Expts}}^{\text{Total}}$ is 5^6 or 15,625. By using Dimensional Analysis to determine the solution function for this example, we reduce the required number of experiments by a factor of 125, thereby greatly reducing the time and cost of defining the function.

The second Dimensional Analysis method we are taught utilizes our knowledge of the conservation laws; i.e., the differential equations, describing the physical or chemical process we are investigating. In most cases, we know these differential equations and we can inscribe them on paper. Unfortunately, we cannot solve them readily. If we could solve them, then we could readily upscale from laboratory-sized equipment to commercial-sized equipment via calculation. The fact that we cannot solve such equations does not, however, mean we cannot use them. We can convert such dimensional differential equations into nondimensional differential equations and, by doing so, identify the terms that control the physical or chemical process under investigation.[9]

Let us consider, again, the isothermal flow of a viscous Newtonian fluid. For constant fluid density and fluid viscosity, the three-dimensional equation of change is

$$\rho\left(\frac{\partial u}{\partial t}\right) + \rho\left(u\frac{\partial u}{\partial x} + v\frac{\partial u}{\partial y} + w\frac{\partial u}{\partial z}\right) = \rho g \cos \alpha_x - \frac{\partial p}{\partial x} + \mu\left(\frac{\partial^2 u}{\partial x^2} + \frac{\partial^2 u}{\partial y^2} + \frac{\partial^2 u}{\partial z^2}\right)$$

where ρ is fluid density (kg/m^3); μ is fluid viscosity (kg/m/s); t is time (s); u is fluid velocity in the x-direction (m/s); v is fluid velocity in the y-direction (m/s); w is fluid velocity in the z-direction (m/s); g is the acceleration of gravity (m/s^2); α_x is the angle at which the fluid flows in the x-direction; p is pressure (kg/s^2 m); and, x, y, and z are directed lengths (m). We could have used pounds, feet, and seconds as our units since the magnitude of the measurement units; i.e., the system of units, does not impact the relationships derived via Dimensional Analysis.[10,11]

If we write the above equation of change without operators and in one dimension, we have

$$\frac{\rho u}{t} + \frac{\rho u^2}{x} = \rho g \quad -\frac{\Delta p}{x} \quad +\frac{\mu u}{x^2}$$
$$\quad 1 \qquad 2 \qquad 3 \qquad 4 \qquad 5$$

Each of the terms in the above equation represents a force. Term 1 is the force required to accelerate a unit volume of fluid during unsteady flow. Term 2 describes the change in momentum across a unit cross-sectional area perpendicular to the direction of flow. Term 3 represents the body forces experienced by a unit volume of fluid. Term 4 is the static pressure gradient within the unit fluid volume. And, lastly, Term 5 describes the reaction of the unit volume of fluid to applied shear force, which is viscous resistance. We can convert the above dimensional equation into a dimensionless equation by dividing each term by Term 5.[11] In other words, divide each force term by the viscous resistance or viscous force. Doing so yields

$$\left(\frac{x^2}{\mu u}\right)\left(\frac{\rho u}{t}\right) \quad + \left(\frac{x^2}{\mu u}\right)\left(\frac{\rho u^2}{x}\right) \quad = \left(\frac{x^2}{\mu u}\right)(\rho g) \quad - \left(\frac{x^2}{\mu u}\right) \quad \left(\frac{\Delta p}{x}\right) + 1$$
$$\qquad\quad 1 \qquad\qquad\qquad 2 \qquad\qquad\qquad 3 \qquad\qquad 4$$

Multiplying Term 3 by one; i.e., by u/u, and multiplying Term 4 by one; i.e., $\rho u/\rho u$, then rearranging each term yields

$$\left(\frac{\rho x u}{\mu}\right)\left(\frac{x}{ut}\right) \quad + \left(\frac{\rho x u}{\mu}\right) \quad = \left(\frac{\rho x u}{\mu}\right)\left(\frac{xg}{u^2}\right) \quad - \left(\frac{\rho x u}{\mu}\right) \quad \left(\frac{\Delta p}{\rho u^2}\right) + 1$$
$$\qquad\quad 1 \qquad\qquad\qquad 2 \qquad\qquad\qquad 3 \qquad\qquad\qquad 4$$

One parenthetical group is common to all the terms in the above equation; it is

$$\frac{\rho x u}{\mu}$$

The above dimensionless parameter is ubiquitous to the analysis of fluid flow. We call it the Reynolds number, in honor of Osborne Reynolds, who began a series of experiments into the nature of fluid flow in 1883. His first set of experiments identified two flow regimes: laminar flow and turbulent flow. Reynolds demonstrated in a second set of experiments conducted during the latter half of the 1880s that flow regime depends upon the above ratio. If $\rho x u/\mu < 2100$, then the flow regime is laminar; if $\rho x u/\mu > 4000$, then the flow regime is turbulent. If $2100 < \rho x u/\mu < 4000$, then the flow regime is unpredictable: it will either be laminar or turbulent, depending upon many variables beyond our control.[12] Note that the Reynolds number is the ratio of inertial forces to viscous forces.[13]

The second parenthetical group in Term 1, x/ut, is a modified Strouhal number.[14,15] The Strouhal number describes oscillating flow mechanisms.

The second parenthetical group in Term 3 is the inverse of the Froude number. The Froude number is

$$\frac{u^2}{xg}$$

which is the ratio of inertial forces to gravitational forces.

The second parenthetical group in Term 4 is the "pressure coefficient" or Euler number. It is the ratio of pressure forces to inertial forces.[16] All these ratios, except the Strouhal number, have a common feature: they are the ratio of a driving force to a resisting force.

Returning to our problem, the last equation above indicates that the solution is the function

$$f\left\{\left(\frac{\rho x u}{\mu}\right), \left(\frac{x}{ut}\right), \left(\frac{xg}{u^2}\right), \left(\frac{\Delta p}{\rho u^2}\right)\right\} = 0$$

Identifying $\frac{\Delta p}{\rho u^2}$ as the dimensionless coefficient allows us to write

$$\kappa g\left\{\left(\frac{\rho x u}{\mu}\right), \left(\frac{x}{ut}\right), \left(\frac{xg}{u^2}\right)\right\} = \frac{\Delta p}{\rho u^2}$$

as the solution to the problem. We must then perform experiments to determine the functional relationship between the dimensionless coefficient and the dimensionless parameters and to determine the value of κ.

Both these methods have problems. With regard to the Method of Indices, the number of unspecified indices grows as the number of constants, parameters, and variables describing the process increases. Performing the necessary algebra to identify the dimensionless parameters quickly becomes difficult and becomes a bookkeeping issue. But, more importantly, we must ensure, after identifying the dimensionless parameters, that they are linearly independent of each other.

With regard to the second method, we must have a complete set of differential equations and, again, we must determine that each of the resulting dimensionless parameters is linearly independent of all others.

What is not taught during these brief overviews of Dimensional Analysis is the concept of dimensions, systems of units and what units represent, and why the Method of Indices works. We must fully understand these points before we can recognize the power of Dimensional Analysis.

And lastly, what chemical engineers need is a procedure for doing Dimensional Analysis that greatly reduces the likelihood of making an algebraic error, that simplifies indices bookkeeping, and that guarantees the linear independence of the resulting dimensionless parameters. This book provides such a procedure.

1.4 SUMMARY

In this chapter, we discussed the historical procedures used by chemical engineers when developing a new process or modifying an existing process. As much as we desire to develop a process using the three conservation laws, we are unable to do so for a variety of reasons. However, we can use Dimensional Analysis and experimentation to define the functional relationships of the variables pertinent to a process. We briefly explained the two most common methods of Dimensional Analysis taught to chemical engineers and we discussed their shortcomings.

We also demonstrated in this chapter how Dimensional Analysis can reduce process development cycle time and costs. This feature of Dimensional Analysis is important in light of the three revolutions that occurred during the second half of the twentieth century. Those revolutions were in transportation, communication, and finance.

NOTES AND REFERENCES

1. E. Nauman, *Chemical Reactor Design, Optimization, and Scaleup*, Second Edition, John Wiley & Sons, New York, NY, 2008, p. 50.

2. V. Skoglund, *Similitude: Theory and Applications*, International Textbook Company, Scranton, PA, 1967, p. 18.

3. More on "dimensional homogeneity" later.

4. A. Porter, *The Method of Dimensions*, Second Edition, Methuen & Co., London, UK, 1943.

5. More on "fundamental dimensions" later.

6. More on M_F and M_W later.

7. R. Pankhurst, *Dimensional Analysis and Scale Factors*, Chapman & Hall, London, UK, 1964, p. 70.

8. P. Grassmann, *Physical Principles of Chemical Engineering*, Pergamon Press, Oxford, UK, p. 507, 1961 (German Edition), 1971 (English Edition).

9. A. Klinkenberg, H. Mooy, *Chemical Engineering Progress*, *44* (1), 17 (1948).

10. R. Johnstone, M. Thring, *Pilot Plants, Models, and Scale-up Methods in Chemical Engineering*, McGraw-Hill, New York, NY, 1957, p. 27.

11. A. Bisio, R. Kabel, *Scaleup of Chemical Processes: Conversion from Laboratory Scale Tests to Successful Commercial Size Design*, John Wiley & Sons, New York, NY, 1985, p. 62.

12. W. Badger, J. Banchero, *Introduction to Chemical Engineering*, McGraw-Hill, New York, NY, 1955, p. 31–34.

13. R. Bird, W. Stewart, E. Lightfoot, *Transport Phenomena*, John Wiley & Sons, New York, NY, 1960, p. 342.

14. R. Pankhurst, *Dimensional Analysis and Scale Factors*, Chapman & Hall, London, UK, 1964, p. 96.

15. D. Boucher, G. Alves, *Chemical and Engineering Progress*, 55 (9), 55 (1959).

16. N. de Nevers, *Fluid Mechanics for Chemical Engineers*, Second Edition, McGraw-Hill, New York, NY, 1991, p. 433–448.

CHAPTER 2

History of Dimensional Analysis

2.1 PRE JOSEPH FOURIER

The concept of dimension is as old as Greek mathematics, but the use of dimension as an analytical tool is relatively modern. Greek mathematics, i.e., geometry, is based on length and dimensionless angle. The Greeks did not consider the implications of dimension since all their mathematical manipulations involved only lengths and angles.

When our earliest ancestors learned to count is unknown, but it surely began shortly after they realized they had fingers and toes. With those fingers and toes, the rules of pure number manipulation came to light. Thus, by the time history began, our ancestors knew how to manage pure numbers; they knew the rules of arithmetic.

With the development of algebra, higher mathematics freed itself from geometry. In algebra, numbers can represent physical quantities, which have dimensions. Attaching dimensional information to numbers negates the rules of arithmetic, unless we observe certain restrictions.[1]

The first to discuss the concept of dimension was Johannes de Mures (c.1290–c.1355), a French philosopher, astronomer, mathematician, and music theorist. He wrote about products and quotients possessing different dimensions. However, his work on products and quotients made no lasting impression on the development of science.

Descartes (1596–1650) may have been the first natural philosopher and mathematician to realize that derived dimensions exist, such as Force.[2] According to Descartes, "[t]he force to which I refer always has two dimensions, and it is not the force that resists (a weight) which has one dimension".[3] However, Descartes was not hindered by mathematical operations that produced dimensionally impossible results. For Descartes, dimensional correctness did not determine the correctness of a given result.

Sir Issac Newton (1642–1727) recognized the concept of derived dimensions: "I call any quantity a genitum which is not made by addition or subtraction of divers parts, but is generated or produced in arithmetic by multiplication, division, or extraction of the root of any term whatsoever ...."[4] Gottfried Leibniz (1646–1716) also recognized the concept of derived dimensions, no doubt to Sir Issac's chagrin: "... action ... is as the product of the mass multiplied by space and velocity, or as the time multiplied by *vis viva*."[4]

The eighteenth century witnessed great advances in analysis of physical phenomena; however, little thought was given to dimensions. Leonhard Euler (1707–1783) was the only natural philosopher and mathematician to make comment on dimensions during that momentous century. In fact, Euler demonstrated a preoccupation about the meaning of physical relationships. In 1736, Euler published *Mechanica* in which he showed that the dimension of n in the equation

$$A \times dv = np \times dx$$

depended on the dimensions of A (mass) and p (force). This observation by Euler indicates that he understood the need for unit homogeneity; that is, the units left of an equal sign must be the same as those units to the right of the same equal sign. Euler further discussed dimensions in his *Theoria motus corporum solidorum seu rigidorum* published in 1755. In this book, Euler devoted a chapter to questions of units and homogeneity. Unfortunately, his writings about dimension made little impression upon the community of mathematicians and natural philosophers of the time.

2.2 POST JOSEPH FOURIER

Little, if any, discussion of dimensions occurred after Euler's *Theoria* until 1822 when Joseph Fourier published the third edition of his *Analytical Theory of Heat*. Fourier makes no mention of dimension in either the first edition of his book, published in 1807, or the second edition, published in 1811. However, in the 1822 edition, Fourier specifically states that any system of units can be used to study a physical process, so long as the chosen system of units is consistent. He also states that mathematical equations used to describe physical processes must demonstrate homogeneity: the units on either side of an equal sign must be the same. Fourier used the concept of homogeneity to check his mathematical manipulations. He clearly states in the 1822 edition that a natural philosopher should use unit homogeneity as a check on his mathematical analysis of a physical process.[4] If the units are not the same on either side of the final equal sign, then the natural philosopher has incorrectly manipulated a mathematical equation occurring earlier in the study—thus, the "birth" of Dimensional Analysis.

While Fourier may have published the basics of Dimensional Analysis and presented the need for unit homogeneity when investigating a physical process, few, if any, pursued his insights. In fact, confusion over dimensions and units persisted through the two middle quarters of the nineteenth century.[5] It was the development of electrical technology and telegraphy, in general, and the trans-Atlantic telegraph cable, in particular, that forced a discussion and review of dimensions and units in the 1860s.[6]

In 1861, the British Association for the Advancement of Science (BAAS) formed a committee to review the various systems of units for electricity and magnetism in use at the time. The committee was also charged with codifying a standard system of units, especially for electrical measurements. This initiative by BAAS was the first effort by engineers and scientists to develop an understanding of units and to establish a standardized system of units.

William Thomson (later Lord Kelvin) and James Clerk Maxwell served on the committee and greatly influenced its program.[7] Maxwell provided the greatest insight into the effort, but he also created the most confusion, confusion that continues to this day. Maxwell realized that physical concepts are quantified by dimensions, e.g., by Length, Mass, and Time. Various sets of dimensions could be grouped and called fundamental dimensions, from which other dimensions, such as Force, Energy, and Power, could be derived. Maxwell suggested identifying fundamental dimensions by brackets; thus, the fundamental dimensions Length, Mass, and Time would be identified as [LMT]. He did not clarify what he meant by his bracket notation; hence, confusion developed with regard to them and still exists about them today. In reality, we should consider them as identifying the procedures to be used when describing a physical concept. Once we have identified the fundamental dimensions for describing a physical concept, then we can develop a set of standards to quantify the chosen dimensions: those standards form a system of units. Many engineers and scientists expended considerable effort during the last quarter of the nineteenth century developing the standards for various systems of units.[8,9]

In 1877, Lord Rayleigh published his *Theory of Sound.* Its index contains an entry entitled "Method of Dimensions."[4] Lord Rayleigh made good use of Dimensional Analysis during his long and fruitful scientific career; however, he never presented a derivation of his method. He simply stated the method could be used as a research tool when investigating physical processes. Lord Rayleigh equated the powers or indices of the dimensions that describe a physical process. His method works well for simple mechanical processes where the number of unknown exponential indices equals the number of equations.[10] For processes involving heat or mass transfer, there will be more unknowns than equations; therefore, the practitioner of Lord Rayleigh's method must assign a value to each of these unknowns, then prove that the assumed unknowns yield an independent set of results. In other words, this method becomes cumbersome and time consuming when applied to complex physical and chemical processes.

While engineers and scientists in Great Britain used Dimensional Analysis without mathematical validation of it, their colleagues on the Continent were investigating the concept of dimension at a more philosophical level. In 1892, A. Vaschy, a French electrical engineer, published a version of what became known as Buckingham's Pi theorem.[11] In the first chapter of his *Theorie de l'Electricite*, published in 1896, Vaschy discusses dimensions, systems of units, and measurements.[4] He presents Buckingham's Pi theorem in modern notation in Chapter One. Unfortunately, the scientific community lost interest in Dimensional Analysis after Vaschy's publication because no reference to Dimensional Analysis occurs until 1911.

In 1911, D. Riabouchinsky published a paper which rediscovered Vaschy's results.[12] Riabouchinsky made this discovery while analyzing data he had generated at the Aerodynamic Institute of Kutchino. He provided a mathematical foundation for Dimensional Analysis and he stated, as did Vaschy, Buckingham's Pi theorem. Riabouchinsky apparently rediscovered the foundation of Dimensional Analysis independent of Vaschy.

In 1914, Richard Tolman and Edgar Buckingham each published an article concerning Dimensional Analysis in the *Physical Review.*[13,14]

Tolman used Dimensional Analysis to investigate Debye's recently published theory of specific heat. Buckingham investigated the foundation of Dimensional Analysis itself. In his paper, he stated, and proved, what became known as Buckingham's Pi theorem, which forms the foundation of Dimensional Analysis. It can be stated as

> If there exists a unique relation $f(A_1, A_2, A_3, ..., A_n) = 0$ among n physical quantities which involve k physical dimensions, then there also exists a relation $\Phi(\pi_1, \pi_2, \pi_3, ..., \pi_n) = 0$ among $(n − k)$ dimensionless products [comprised] of A's.[15]

Buckingham's proof of this theorem was original and independent of Vaschy's and Riabouchinsky's proofs. All of them placed Dimensional Analysis on a mathematical foundation.

In 1922, Percy Bridgman published a small book entitled *Dimensional Analysis*.[16] Bridgman presented an overview of Dimensional Analysis to date and provided a proof confirming Rayleigh's Method of Indices. Sporadic discussion of dimension, systems of units, and Dimensional Analysis occurred during the 1930s and 1940s. In 1951, Henry Langhaar published *Dimensional Analysis and Theory of Models* in which he formulated Dimensional Analysis in a matrix format.[17] Langhaar proved the concept of dimensional homogeneity in that book.

Before the advent of digital computing, using matrices for Dimensional Analysis was not much easier than using Rayleigh's Method of Indices. However, with the ever increasing power of computers and with the availability of ever increasingly user-friendly software, the popularity of the matrix formulation of Dimensional Analysis has grown, as attested by the publication of Thomas Szirtes' *Applied Dimensional Analysis and Modeling* and Marko Zlokarnik's *Scale-up in Chemical Engineering*.[18,19] The former book presents the matrix algebra required by Dimensional Analysis and mainly applies the matrix format to mechanical and structural engineering examples. The latter book has less mathematical theory; its emphasis is apparent in its title.

2.3 SUMMARY

Dimension and its importance was not well understood until 1822 when Joseph Fourier stated that the units left and right of an equality sign should be the same for equations containing physical information. When manipulating pure numbers, this requirement does not arise—a number is a number. But, in the physical sciences, where equations contain physical information about Nature, the units on either side of an equality sign do matter. It was not until the electrical revolution of the mid-nineteenth century that engineers and scientists became interested in dimension and units. At that time, they did not realize that dimension and units are not the same concept. In fact, the two concepts are confused and used interchangeably today.

At the turn of the twentieth century, some engineers and scientists began to realize that dimension and units are different concepts and that dimension could be used as a separate concept for gaining a deeper understanding of the relationships underlying physical processes; thus, the birth of Dimensional Analysis.

NOTES AND REFERENCES

1. J. Hunsaker, B. Rightmire, *Engineering Applications of Fluid Mechanics*, McGraw-Hill, New York, NY, 1947, Chapter 7.

2. A derived dimension is a dimension formed from a combination of the fundamental dimensions (see below). For example, if Length, Mass, and Time are fundamental dimensions, then Force (LM/T^2) is a derived dimension.

3. R. Dugas, *A History of Mechanics*, Dover Publications, New York, NY, 1988.

4. E. Macagno, *Journal of the Franklin Institute*, 292 (6), 391 (1971).

5. B. Mahon, *The Man Who Changed Everything: The Life of James Clerk Maxwell*, John Wiley & Sons, Chichester, UK, 2003, Chapter 8.

6. J. Gordon, A Thread, *Across the Ocean: The Heroic Story of the Transatlantic Cable*, Walker Publishing Company, Inc, New York, NY, 2002.

7. D. Lindley, *Degrees Kelvin: A Tale of Genius, Inventions, and Tragedy*, John Henry Press, Washington, DC, 2004, pp. 142–153.

8. A. Klinkenberg, *Chemical Engineering Science*, 4,, 130–140, 167–177.

9. F. Civan, *Chemical Engineering Progress*, 43–49 (February 2013).

10. A. Porter, *The Method of Dimensions*, Second Edition, Methuen and Co. Ltd, London, UK, 1943.

11. A. Vaschy, *Annales Telegraphiques*, 19, 25–28 (1892).

12. D. Riabouchinsky, *L'Aerophile*, 19, 407–408 (1911).

13. R. Tolman, *Physical Review*, 4, 145–153 (1914).

14. E. Buckingham, *Physical Review*, 4, 345–376 (1914).

15. S. Corrsin, *American Journal of Physics*, 19, 180–181 (1951).

16. P. Bridgman, *Dimensional Analysis*, Yale University Press, New Haven, CT, 1922.

17. H. Langhaar, *Dimensional Analysis and Theory of Models*, John Wiley & Sons, New York, NY, 1951.

18. T. Szirtes, *Applied Dimensional Analysis and Modeling*, Second Edition, Butterworth–Heinemann, Burlington, MA, 2007.

19. M. Zlokarnik, *Scale-Up in Chemical Engineering*, Second Edition, Wiley-VCH Verlag GmbH & Co.KGaA, Weinheim, Germany, 2006.

CHAPTER 3

Dimensions and Systems of Units

3.1 PHYSICAL CONCEPT AND PHYSICAL QUANTITY

Dimensional Analysis is based on the concept of physical quantity. When we are young, we are taught to manipulate pure numbers. First, we are taught to add and subtract; then, we are taught to multiply and divide numbers. During our Junior High and Senior High school years, we are introduced to science and taught that science depends upon quantifying physical concepts and manipulating the numbers resulting from that quantification. However, we are not told what those quantifying numbers represent. We are only told that we can sum apples to obtain apples, but we cannot sum apples and oranges to obtain "orples." Our teachers do not prove these statements, and, in fact, they cannot be proven without first defining physical concept and physical quantity, which our teachers do not discuss.

But, what is physical concept? A physical concept describes a sensory perception. For example, we experience Length as height, depth, and breadth. We experience Mass as weight, which we describe as "light" or "heavy" relative to a reference, which could be ourselves. We experience Time in a variety of ways. First, we experience time cycles: the sun rising, the sun setting; the moon full, the moon absent; the tide rising, the tide falling. Second, we experience time as "short" or "long" relative to a reference event. And, lastly, as we grow older, we experience Time as a succession of events. Force we experience as impact: one impact is less than or greater than a previous impact. Temperature we experience as "cold" and "hot": river ice is cold; fire is hot.

To quantify a physical concept, we first determine a descriptor that best characterizes it, that produces a valid physical quantity. Thus, a physical quantity represents a quantitative description of a physical concept. In other words, a physical concept attains meaning only if its descriptors can be measured. Those descriptors are dimensions.

At some point in time, one of our ancestors, stood and bumped his or her head against the cave roof. Thus, the physical concept of height was born. That physical concept became a physical quantity when it was defined as the length. With the advent of agriculture, our ancestors realized the need to describe garden plots in order to trade them or estimate their food content. Visually, they could see that a garden plot has two lengths, what we call width and breadth; hence, the physical concept of area. To quantify the concept of area, they most likely stepped the width from one corner to an adjoining corner, then took a step along the breadth, turned, and stepped the width again, counting each step as it was taken. They continued this procedure until they had traversed the entire garden plot. For a 50 step width and a 100 step breadth garden plot, they counted a total of 5000 steps. At some point, someone realized that a 50 step width and a 100 step breadth always produced 5000 steps, thereby completing the quantification of area and introducing the mathematical concept of multiplication.

Our ancestors did not stop developing physical concepts once they understood length, area, volume, mass, temperature, and time. With each new discovery in Nature, a new physical concept had to be developed, then converted through quantification into a physical quantity. Eventually, our ancestors developed the physical concept of force and energy, and, most recently, all the physical concepts associated with particle physics, such as "strangeness." Upon quantification, all these physical concepts became physical quantities.

James Clerk Maxwell suggested, not clearly, that we designate the physical quantity of a physical concept with brackets. For example, consider length: the physical quantity of that physical concept is $[L^1]$; the physical quantity of the physical concept we call area is $[L^2]$; and, the physical quantity of the physical concept we call volume is $[L^3]$. The power indices on L actually designate "dimension."[1] Thus, the index 1, or unit power, designates lines; the index 2, or square power, designates planes or areas; and, the index 3 designates cubes or volumes.

Dimensions come in two flavors: fundamental and derived. Fundamental dimensions form a basic set of quantifiers for describing a physical concept. The quantifiers of such a set must be independent of each other. For example, we can choose length [L], mass [M], and time [T] as our set of fundamental dimensions. Or, we could choose length [L], force [F], and time [T] as our set of fundamental dimensions. Each choice forms a set of independent fundamental dimensions.

Derived dimensions arise from our study of Nature: many observations of Nature require us to combine fundamental dimensions in order to quantify them. Watching clouds scurry across the sky forced one of our ancestors to recognize the physical concept of speed, which is described by the physical quantity of [L]/[T] or [LT^{-1}]. [LT^{-1}] is a derived dimension. Using the fundamental dimensions [LMT], we would derive the force [F] via Newton's Second Law, namely

$$F = ma$$
$$F[=]ML/T^2 = MLT^{-2}$$

where [=] implies our equation uses dimension notation rather than mathematical notation. Using the fundamental dimensions [FLT], we would derive [M] using the same law, in other words

$$F = ma$$
$$F/a = m$$
$$M[=]FT^2/L = L^{-1}FT^2$$

We choose the LMT fundamental dimension set for situations involving dynamics, and we choose the LFT fundamental dimension set for situations involving statics. Table 3.1 gives various fundamental dimension sets and their derived dimensions.

Note that neither fundamental dimensions nor derived dimensions involve magnitudes. Dimension describes the nature of a physical concept without introducing magnitude.[1(p20)]

Table 3.1 Various Fundamental Dimensions

Physical Quantity	Fundamental Dimensions					
	LMT	LFT	LET	LMTθ	LFTθ	LQTθ
Area	L^2	L^2	L^2	L^2	L^2	L^2
Volume	L^3	L^3	L^3	L^3	L^3	L^3
Velocity	LT^{-1}	LT^{-1}	LT^{-1}	LT^{-1}	LT^{-1}	LT^{-1}
Acceleration	LT^{-2}	LT^{-2}	LT^{-2}	LT^{-2}	LT^{-2}	LT^{-2}
Force, weight	LMT^{-2}	F	$L^{-1}E$	LMT^{-2}	F	$L^{-1}E$
Density	$L^{-3}M$	$L^{-3}F$	$L^{-4}E$	$L^{-3}M$	$L^{-3}F$	$L^{-4}E$
Energy	L^2MT^{-2}	LF	E	L^2MT^{-2}	LF	E
Power	L^2MT^{-3}	LFT^{-1}	ET^{-1}	L^2MT^{-3}	LFT^{-1}	ET^{-1}
Mass	M	$L^{-1}FT^2$	$L^{-2}ET^2$	M	$L^{-1}FT^2$	$L^{-2}ET^2$
Entropy				$L^2MT^{-2}\theta^{-1}$	$LF\theta^{-1}$	$E\theta^{-1}$
Thermal conductivity				$LMT^{-3}\theta^{-1}$	$T^{-1}F\theta^{-1}$	$L^{-1}T^{-1}E\theta^{-1}$

What dimension does is associate a rule or procedure, i.e., a standard measuring procedure, with a physical concept. We call the standard measuring procedure a "descriptor": it describes what needs measuring to transform a physical concept into a physical magnitude.

3.2 PHYSICAL MAGNITUDE

The quantitative description of a physical concept involves a comparison with a standard measure through the use of a standard procedure, the comparison being either a fraction or a multiple of the standard measure. Thus, a number will be associated with the physical quantity, which transforms it into a physical magnitude. The physical magnitude has an associated error since each standard measure includes a measurement error. If we denote the physical concept as α and its descriptor as $[\psi]$, then we can denote the physical quantity as $\alpha[\psi]$. A physical magnitude is any physical quantity to which we can affix a numerical value. Thus, physical magnitudes are[2]

$$\alpha[\psi] = \{\text{numerical value}\} \times \{\text{standard unit of measure}\} \pm \text{error}$$

Note that we can establish a physical concept for taste and bouquet and physical quantities for both, with dimension $[\tau]$ for taste and dimension $[\sigma]$ for smell. However, we cannot convert these physical quantities into physical magnitudes because we are unable to establish a standard unit for either, which means we are unable to assign a numerical value to either physical quantity.[3]

It is the above definition that leads us to develop systems of units. The descriptor $[\psi]$ tells us what dimensions require measuring procedures with regard to the physical concept α. For example, if α is a length of rope, then $[\psi]$ is [L], which tells us we require an approved procedure for measuring length and we require a standard unit of measure to quantify length. The procedure may be stated as: stretch the rope flat on level ground; one person holds one end while a second person holds the other end; a third person then lays a standardized piece of straight wood, say the average length of a Roman soldier's foot, next to the rope, aligning one end of the stick with one end of the rope; upend the stick repeatedly along the length of the rope; count each upending of the stick; stop counting when the stick reaches the other end of the rope; estimate what portion of the stick aligns against the rope's end. Let us say we followed this procedure and upended the stick nine times. On the tenth upending, the stick extended beyond the rope. Viewing the rope and stick, we estimate the rope's end aligns at the mid-point of the stick. Thus, we conclude the rope is 9½ stick lengths. If we define the stick length as a "foot," then the rope length is 9½ ft. If the stick is graduated evenly with 11 marks, we then write, letting $\alpha = \ell$,

$$\ell[L] = 9\frac{1}{2} \text{ ft} \pm 1/12 \text{ ft}$$

Thus, we have converted a physical concept into a physical quantity, then into a physical magnitude.

We write the physical quantity and the physical magnitude in the above format for a reason: it separates the magnitude from the units. The magnitude is a fractional quantity or it is a multiple of the unit standard. In other words, the magnitude is a number and, as such, subject to mathematical manipulation. The unit standard identifies the units we use to establish the magnitude. For example, consider a runner who sprints a quarter mile (440 yards or 1320 ft) in 60 s. We want to know how fast he or she ran 1320 ft in 60 s. The physical concept, α, is speed, symbolized as S, and its physical quantity is $[LT^{-1}]$. Its physical magnitude is

$$S[LT^{-1}] = (1320) \times (1 \text{ ft}) \div (60) \times (1 \text{ s})$$

Thus

$$S[LT^{-1}] = \left(\frac{1320}{60}\right) \times \left(\frac{1 \text{ ft}}{1 \text{ s}}\right)$$

Performing the indicated mathematical operation on the pure number portion of the physical magnitude yields

$$S[LT^{-1}] = 22 \times \left(\frac{1 \text{ ft}}{1 \text{ s}}\right)$$

Writing the above in a more conventional form gives us

$$S[LT^{-1}] = 22 \text{ ft/s}$$

The format of the physical magnitude is important because we cannot divide 1320 ft by 60 s. However, we can divide the magnitude of the foot standard unit by the magnitude of the second standard unit. Scientists debated whether units were mathematical entities or not for many decades. They eventually decided units are not mathematical entities; however, their magnitudes are mathematical entities because they arise from counting the standard units required by the measurement.

Consider a second example: one of our ancestors decides that his or her garden plot is best described by the physical concept area. Our ancestor plans to trade garden plots with a neighbor, but our ancestor needs to quantify the two garden plots. Our ancestor, therefore, takes his or her stick, which is the length of the average Roman soldier's foot, and measures the width and breadth of his garden plot per the above length instructions. The width is 100 stick upendings and the breadth is 200 stick upendings. The physical concept is area, $\mathscr{A}$, and the physical quantity is $\mathscr{A}[L^2]$, therefore

$$\mathscr{A}[L^2] = (100) \times (1 \text{ stick}) \times (200) \times (1 \text{ stick})$$
$$\mathscr{A}[L^2] = (100) \times (200) \times (1 \text{ stick}) \times (1 \text{ stick})$$
$$\mathscr{A}[L^2] = (20,000) \times (1 \text{ stick}) \times (1 \text{ stick}) = (20,000) \times (1 \text{ stick})^2$$

Note that we cannot physically combine; i.e., multiply, two sticks, each a foot long, but we can mathematically manipulate the multiples of each stick since the multiples are pure numbers. However, for convenience, we can change the notation of our standard unit to match the dimension of our physical quantity. In other words

$$\mathscr{A}[L^2] = (20,000) \times (1 \text{ stick}) \times (1 \text{ stick}) = (20,000) \times (1 \text{ stick})^2$$

You may wonder why we belabor this point. It is simple: we always confuse dimension and unit. We must remember that dimension defines a physical quantity and stipulates the procedures required to transform that physical quantity into a physical magnitude. On the other hand, a standard unit provides a scale with which to obtain the fraction of or multiple of the standard unit that comprises the physical magnitude.

3.3 SYSTEMS OF UNITS

A unit is a specified quantity for a given dimension; thus, it allows us to create physical magnitudes. We can establish, in any fashion, a unit to determine the magnitude of a specified dimension, just so long as we follow the approved procedure for using that unit. For example, the standard unit of length the scientific community established in 1889 and used until 1960 was the International Meter, defined as the length between two engraved lines on a bar of 90% platinum, 10% irid-ium. After 1960, the scientific community defined the standard unit of length as 1,650,763.73 vacuum wavelengths of 6058 Å light emitted by a Krypton-86 discharge tube. We call this length the optical meter. The accuracy for the platinum−iridium bar is $1/(1 \times 10^6)$, whereas the optical meter has an accuracy of $1/(1 \times 10^8)$.[1(pp40−43)] The scientific community redefined the optical meter in 1983. It is now the distance traveled by light in vacuum in 1/299,792,458 s. ASTM International provides a library of procedures for specifying standard units and for using standard units for measurement.

Units can be extensive or intensive. Extensive units sum to become larger. Length has an extensive unit, e.g., 5 ft of rope may be one piece or it may be five, 1 ft rope segments joined together. Mass has an exten-sive unit. On the other hand, intensive units do not sum to become larger, e.g., temperature, density, and viscosity have intensive units.[4−6]

Systems of standard units, or simply, systems of units are ubiqui-tous in science and engineering. Historically, each scientific and engi-neering discipline developed a system of units that best matched their needs. Each of these systems of units is valid, so long as their practi-tioners use them properly and correctly. This chaotic democracy requires us to know the foundations of each system of units, and it requires an inordinate number of conversion factors.[7] The need for these conversion factors is disappearing as more of the global commu-nity adopts the International System (SI) of Units. Table 3.2 presents the SI fundamental dimensions and the SI system of units. However, the United States and some other countries still use the English Engineering system of units and the American Engineering system of units. Also, much of the scientific and engineering literature, especially prior to the

Table 3.2 SI Fundamental Dimensions			
Dimension	Symbol	Unit	Unit Symbol
Length	L	Meter	m
Mass	M	Kilogram	kg
Time	T	Second	s
Electric current	I	Ampere	A
Thermodynamic temperature	Θ	Kelvin	K
Amount of substance	N	Mole	mol
Luminous intensity	Cd	Candela	cd

Table 3.3 English Engineering Fundamental Dimensions and Units

Dimension	Symbol	Old English Units	New English Units	English Gravitational Units	English Dynamic Units
Length	L	Foot	Foot	Foot	Foot
Mass	M	Pound mass (lb)	Pound mass (lb$_M$)	Slug	Pound (lb)
Time	T	Second or hour	Second or hour	Second	Second
Force	F	Pound force (lb)	Pound force (lb$_F$)	Pound	Poundal
Thermodynamic temperature	Θ	°F	°C	°R	°C

mid-1960s, uses systems of units other than metric or SI units. Table 3.3 displays these systems of units.

Any number of systems of units can be devised, so long as the fundamental dimensions upon which they are founded are independent of each other. The simplest such systems of units are based on three fundamental dimensions. They are Length, Mass, and Time or Length, Force, and Time. Newton's Second Law links these fundamental dimensions, namely, for Length, Mass, and Time

$$F = ma = [LMT^{-2}]$$

and for Length, Force, and Time

$$m = \frac{F}{a} = [L^{-1}FT^2]$$

In the former system of units, one kilogram (kg) of mass accelerated at one meter per second squared (m/s^2) experiences 9.81 kilogram-meter/time squared (kg m/s^2) of force. In the latter system of units, 9.81 kg m/s^2 of force imparts an acceleration of 1 m/s^2 to 1 kg of mass. Neither of these systems of units requires a dimensional constant. We call such systems of units "absolute."[1(p25)] We refer to absolute systems of units based on Length, Mass, and Time as dynamic systems of units.[8] Physicists and engineers designing mechanisms with moving parts use dynamic systems of units.

We call systems of units based on Length, Force, and Time static systems of units. We sometimes call them gravitational systems of units.[8] Structural engineers, civil engineers, and engineers designing stationary equipment prefer to use static systems of units. Thus, in dynamic and gravitational systems of units, Newton's Second Law does not require a constant; it is, simply

$$F = ma$$

Before the metric system of units gained wide popularity, Great Britain used foot, pound, and second as its system of units. In this system of units, 1 lb force imparts 32 ft/s^2 acceleration to 1 lb mass. This system of units was widely used until the mid-1960s. At the same time, mechanical engineers in Great Britain wanted to use the pound to define force while chemists, and later chemical engineers, wanted to use the pound for mass.[7] To accommodate both groups, engineers and scientists started using a system of units with four fundamental dimensions, those dimensions being Length, Mass, Force, and Time.

We call systems of units using fundamental dimensions of Length, Mass, Force, and Time "engineering systems of units." An enormous amount of confusion exists in the chemical processing industry because of these systems of units. The confusion arises because we give Force and Mass the same unit identification, namely, kilogram or pound. Also, when publishing results, scientists and engineers tend not to specify which system of units they are using in their memo, paper, or report. Thus, we, the readers, must determine whether the author is using an engineering, gravitational, or dynamic system of units. Adoption of the SI system of units solves this problem for future scientists and engineers; however, much of our technical information is historic. Thus, when we read a published paper or report, we must determine which system of units the author used in order to correctly implement his or her data and conclusions.

Historically, the four most common engineering systems of units are

1. Old English Engineering[9]
2. New English Engineering
3. English Gravitational (Static)
4. English Dynamic (Absolute).

Table 3.2 presents these systems of units, as well as the fundamental dimensions defining each unit system. While we may apply the above names as though they were commonly used, they were not and are not. Each system of units has been identified differently when authors have written about units. For example, both the Old English and New English Engineering systems of units have been identified as the "American Engineering" system of units. Also, some authors have identified the English Gravitational system of units as the "British Engineering" system of units.[7] We identify the above systems of units collectively as "mechanical" systems of units because all of mechanics can be described using them.

The Old English Engineering system of units uses pound (lb) for force, pound (lb) for mass, foot (ft) for length, and second or hour (s or h) for time. Note that lb force and lb mass are not stipulated in this system of units. You have to be an astute, careful reader to distinguish which pound an author is discussing. This system of units has caused much confusion among scientists and engineers.

To alleviate that confusion, engineers proposed the New English Engineering system of units, in which pound force (lb_F) and pound mass (lb_M) are separately identified. This notation reduces the confusion inherent in identifying two units with the same symbol.

This identification of lb_F and lb_M brings us to the underlying cause of confusion concerning engineering systems of units, namely, force is defined and not derived. Newton's Second Law states that

$$F = ma$$

In other words, we have two independent variables and one dependent variable. We can specify the dimensions for any two variables, thereby allowing the third variable to become the dependent variable whose dimensions are derived from the dimensions of the two independent variables. However, when we specify the dimensions of all three variables, we over-specify the equation. Thus, in the Old English Engineering system of units, Newton's Second Law has units of

$$lb = lb(ft/s^2)$$

which is highly confusing. In the New English Engineering system of units, Newton's Second Law has units of

$$lb_F = lb_M(ft/s^2)$$

From the above "equations," we immediately see that we need a dimensional constant to achieve dimensional homogeneity. The dimensional constant is the gravitational constant g_O and g_C, respectively.

Consider the gravitational constant g_C: what is it? We define Newton's Second Law by measuring the acceleration that 1 lb force imparts to 1 lb mass during free fall. We stipulate that free fall occurs at 45° latitude and at sea level. Free fall acceleration thus defined is 32 ft/s². Newton's Second Law, therefore, becomes

$$1 \text{ lb}_F = 1 \text{ lb}_M (32 \text{ ft/s}^2)$$

To balance this equation, we see that g_C must be 32 lb_M ft/lb_F s². Newton's Second Law thus becomes

$$(32 \text{ lb}_M \text{ ft/lb}_F \text{ s}^2)(1 \text{ lb}_F) = (1 \text{ lb}_M)(32 \text{ ft/s}^2)$$

In mathematical notation, we write the above equation as

$$g_C F = ma$$

What about g_O? In the Old English Engineering system of units, we write Newton's Second Law as

$$\text{lb} = \text{lb}(32 \text{ ft/s}^2)$$

g_O is measured using the same stipulations used to measure g_C. It is 32 ft/s². Thus, for the Old English Engineering system of units, Newton's Second Law is

$$(32 \text{ ft/s}^2)\text{lb} = \text{lb}(32 \text{ ft/s}^2)$$

and, in mathematical notation it is

$$g_O F = ma$$

In the English Gravitational system of units, the fundamental dimensions are length in feet (ft), force (lb or lb_F), and time in seconds or hours (s or h). In this system of units, a 1 lb force imparts an acceleration of 32 ft/s² to a mass of one "slug." Newton's Second Law is then

$$1 \text{ lb} = (1 \text{ slug})(32 \text{ ft/s}^2)$$

Thus

$$1 \text{ slug} = (1/32) \text{ lb s}^2/\text{ft}$$

In the English Dynamic system of units, 1 lb of mass accelerated at 32 ft/s^2 experiences a force of 1 poundal. From Newton's Second Law, 1 poundal is

$$1 \text{ poundal} = (1 \text{ lb})(32 \text{ ft}/s^2) = 32 \text{ lb ft}/s^2$$

Thus, the state of confusion concerning units.

3.4 HEAT AND TEMPERATURE

We know hotness and coldness exist because if we sit or stand near a campfire we become uncomfortable. We say we get hot. On the other hand, if during winter, we fall through the ice covering a lake, we become, again, uncomfortable. We say we are cold. Over the centuries, we developed instruments to measure how hot is hot and how cold is cold. What these instruments measure is temperature, which we have defined by various scales.

We all know that if we place a kilogram of water on a block of hot steel, the temperature of the water increases with time. Thus, something moved from the hot steel block to the water. With considerable effort, scientists during the nineteenth century identified that "something" as the movement of heat. Eventually, scientists and engineers realized that heat was the movement of energy from one location to another location and that temperature provided one means of determining the direction of heat movement. While temperature is related to energy, we generally consider it an independent dimension.

Heat can perform work or mechanical work can "create" heat; for example, if we place a stirrer, powered by a falling weight in a kilogram of water, the temperature of the water increases as the weight falls, provided the walls of the vessel are adiabatic.[10] In this case, mechanical work "created" heat, as indicated by the temperature rise of the water. On the other hand, consider a cylinder of gas sitting on a hot plate. The cylinder is filled with gas and fitted with a moveable piston. As heat moves from the hot plate to the cylinder, the gas expands, thereby raising the piston. In other words, heat does work. In this case, we define the amount of heat transferred from the hot plate to the gas in the cylinder as calories. In both cases, the amount of energy transferred is calculated as

$$q = mC_P \Delta K$$

where q is energy (m^2 kg/s^2); m is mass (kg); C_P is specific heat capacity at constant pressure (m^2/s^2 K); and K is temperature. q can also be calculated in calories, in which case, m is mass (kg); C_P is specific thermal capacity (calories/K); and K is temperature. The former calculation yields $q_{Mechanical}$ and the latter calculation yields $q_{Thermal}$. Therefore, since

$$q_{Thermal} \propto q_{Mechanical}$$

we need a dimensional constant to convert the proportionality to an equality. That dimensional constant is J_C, the mechanical equivalent of heat, which is 1 calorie = 4.184 J.[11,12] Thus, the above proportionality becomes

$$q_{Thermal} = J_C q_{Mechanical}$$

In summary, energy movement, i.e., heat movement, requires a fourth fundamental dimension. That fourth dimension can be either temperature [θ] or energy as heat [E]. In this book, we will use temperature to indicate energy transference.

3.5 CHEMICAL CHANGE

Chemical change occurs, naturally, without our intervention, and intentionally, with our intervention. We perceive chemical change as rust forming on the under-carriage of automobiles, as butter going rancid, as fabric color fading, and myriad other ways. Therefore, we need a fifth fundamental dimension for the description of chemical change. The physical concept for chemical change is amount of reacting substance and its physical quantity is moles [N]. In the SI system of units, the mole determines the physical magnitude of the chemical change. For chemical change, the fundamental dimensions are length, mass or force, time, temperature or energy, and the amount of reacting substance. Thus, four groups of fundamental dimensions can be used to describe chemical change. They are as follows:

- [L, M, T, θ, N]
- [L, M, T, E, N]
- [L, F, T, θ, N]
- [L, F, T, E, N].

So, the question is: which of these four sets of fundamental dimensions do we use? Answer: our first criterion is ease of calculation; our second criterion is to minimize the number of dimensional constants in the calculation. Therefore, we would, first, determine which variables are changing in the system under investigation and choose the set of fundamental dimensions which provide the simplest procedures for measuring the extent of change with respect to each variable. Second, we would choose the set of fundamental dimensions that allow us to perform our calculations with the fewest number of dimensional constants. For example, in most chemical processes, length, expressed as volume, and mass change with time. Also, the amount of substance reacting changes with time. Thus, we would choose L, M, T, and N as fundamental dimensions. If we choose L, F, T, and N, then, in the English Engineering system of units, our mass unit would be "slugs," not a unit producing simple calculations. To avoid the slug, we could use an engineering system of units, such as L, M, F, and N, but then we require a dimensional constant, g_C, to relate mass and force, thereby adding another constant to our calculations. Energy generated or consumed also changes with time in a chemical process. In this case, our choice for fundamental dimension is temperature or energy. But, if we measure energy in calories or BTUs, then we require a dimensional constant, J_C, to convert calories or BTUs into Joules (kg m^2/s^2). Thus, chemical engineers nearly always choose the first set of fundamental dimensions, above, when describing a chemical process.

Table 3.2 gives the fundamental dimensions upon which the SI system of units are based. Table 3.3 provides a survey of various sets of fundamental dimensions. Table 3.3 does not include all the potential sets of fundamental dimensions.

3.6 SUMMARY

In this chapter, we discussed how we form physical concepts from our sense perceptions and how we have learned to convert physical concepts into physical quantities, for which we had to develop an understanding of dimension. We distinguished two types of dimension: fundamental and derived. We then discussed how dimensions led to systems of units, which we need when forming physical magnitudes. We require physical magnitudes in order to apply mathematical logic to equations containing physical content. We concluded with a discussion of the five fundamental dimensions most commonly employed by chemical engineers.

NOTES AND REFERENCES

1. R. Pankhurst, *Dimensional Analysis and Scale Factors*, Chapman & Hall, London, UK, 1964, p. 13.

2. H. Hornung, *Dimensional Analysis: Examples of the Use of Symmetry*, Dover Publications, Mineola, NY, 2006, p. 1.

3. D. Ipsen, *Units, Dimensions, and Dimensionless Numbers*, McGraw-Hill, New York, NY, 1960, Chapter 2.

4. R. Tolman, *Physical Review*, *9*, 237 (1917).

5. G. Lewis, M. Randall, *Thermodynamics and Free Energy of Chemical Substances*, McGraw-Hill, New York, NY, 1923, p. 13.

6. K. Denbigh, *The Principles of Chemical Equilibrium*, Third Edition, Cambridge University Press, Cambridge, UK, 1978, p. 7.

7. A. Klinkenberg, *Industrial and Engineering Chemistry*, *61* (4), 53 (1969).

8. A. Klinkenberg, *Chemical Engineering Science*, *4*, 180 (1955).

9. Not to be confused with the language "Old English." In our case, Old English Engineering means units devised during the nineteenth century and used well into the mid-twentieth century.

10. K. Denbigh, *The Principles of Chemical Equilibrium with Applications in Chemistry and Chemical Engineering*, Third Edition, Cambridge University Press, Cambridge, UK, pp 15–18.

11. J. Joule, *Philosophical Magazine*, *31*, 173 (1847).

12. J. Joule, *Philosophical Magazine*, *35*, 533 (1849).

CHAPTER 4

Foundation of Dimensional Analysis

4.1 INTRODUCTION

Equations come in two varieties: mathematical and physical. Mathematical equations involve numbers that have no physical content, i.e., they involve pure numbers. We explore the relationships between pure numbers using the logic and rules of mathematics. We learn a fair number of these relationships during our mathematical preparation for an engineering career.

Scientists and engineers use physical equations. Physical equations are developed from experimental data and observation. They balance one set of physical magnitudes against another set of physical magnitudes via the equality sign of mathematics. The law for the conservation of energy is a good example of a physical equation. It was developed during the mid-nineteenth century through the effort of many scientists and engineers. For a flowing fluid, the physical concept for the mechanical conservation of energy is

Energy due to Applied Pressure	+	Kinetic Energy	+	Potential Energy	+	Energy Loss due to Friction	=	Shaft Work per Unit Mass

The physical equation for the conservation of energy for a flowing fluid, in the English Engineering system of units, is

$$\int \frac{\mathrm{d}P}{\rho} + \Delta \left(\frac{\langle u \rangle^2}{2\alpha\, g_C} \right) + \left(\frac{g}{g_C} \right) \Delta z + F = -\frac{W_S}{m}$$

where P is pressure $[L^{-2}F]$; ρ is fluid density $[L^{-3}M]$; u is fluid velocity $[LT^{-1}]$—the bracket $\langle\ \rangle$ indicates an averaged value; α [dimensionless] describes the fluid flow profile within the conduit bounding the flowing fluid; g_C is the gravitational constant $[LMF^{-1}T^{-2}]$; g is gravitational acceleration $[LT^{-2}]$; z is height above the datum plane $[L]$; F is net frictional loss due to fluid flow [LFM^{-1}]; W_S is shaft work [LF]; and m is mass [M]. Thus, the dimension for each term is [LFM^{-1}].

4.2 DEVELOPING DIMENSIONAL ANALYSIS

Since physical equations contain physical magnitudes, they must by necessity contain physical content. They contain physical content because physical magnitudes arise from physical quantities, which in turn arise from our perceptions. Therefore, when we write a physical equation, we are, in essence, writing an equation that balances physical quantities $\alpha[\Psi]$ through the use of an equality sign. Thus, we arrive at the first "axiom" of Dimensional Analysis.

> Axiom 1
>   The numerical equality of a physical equation exists only when the physical magnitudes of that particular physical equation are similar, i.e., have the same units, which means the dimensions of the underlying physical quantities $\alpha[\Psi]$ are similar.[1]

In other words, a valid physical equation is dimensionally homogeneous; that is, all its terms have the same dimensions and units.

All engineers and scientists learn this axiom upon their introduction to the study of Nature. We are told upon writing and solving our first physical equation that the individual terms of the given physical equation must have the same dimensions, i.e., units. We are also told that the dimensions and units of our calculated result must agree with the dimensions and units of the individual terms of the physical equation. For example, consider the physical equation

$$W = X - Y + Z$$

We can only calculate W if X, Y, and Z have the same dimensions and units. If X, Y, and Z each represent a physical magnitude of apples, then we can add and subtract them to obtain W, which will be a physical magnitude of apples. If X and Z have apple dimension and Y has orange dimension, then the above expression ceases to be a physical equation; it becomes meaningless from an engineering or scientific viewpoint. Nonhomogeneous expressions do not contain physical information, thus they are not physical equations. The classic example of a nonhomogeneous expression is

$$s + v = \frac{1}{2}at^2 + at$$

where s is distance [L]; v is velocity [LT^{-1}]; a is acceleration [LT^{-2}]; and t is time [T].[1(pp18–20),2] Writing this expression in dimensional terms gives us

$$[L] + [LT^{-1}] = [LT^{-2}][T^2] + [LT^{-2}][T]$$

which yields, upon simplification

$$[L] + [LT^{-1}] = [L] + [LT^{-1}]$$

This last expression contains information, but that information does not describe a relationship between the left- and right-side of the equality sign. No such relationship exists because the dimensions of the individual terms of the expression are mismatched. We frequently encounter nonhomogeneous expressions during our professional careers. Such expressions generally correlate, statistically, a product property to a process variable. In other words, the correlation describes a coincidence, not a cause and effect. Many such correlations exist in the polymer industry. Unfortunately, each such correlation is valid only for a given product from a particular production plant, which means the correlation possesses no physical information for another product or a different production plant.

We classify homogeneous physical equations as "restricted" and as "general." An example of a restricted equation is

$$s = (16.1) \, t^2$$

which describes the distance s [L] traversed by a free-falling object in time t [T]. Dimensionally, the above expression is

$$[L] = [T^2]$$

which makes it nonhomogeneous. However, we know, in certain situations, that it contains valid physical information. For this expression to be true, the coefficient 16.1 must have dimensions [LT^{-2}]. It, therefore, is not unreasonable for us to assume

$$16.1 = \frac{1}{2}g_O$$

where g_O is 32.2 ft/s^2 in the Old English Engineering system of units. Hence

$$s = (16.1) \, t^2$$

is a valid physical equation so long as the coefficient is a dimensional constant with Old English Engineering units. If this condition is true, the above expression becomes a restricted homogeneous physical equation. However, the above expression is not a physical equation if we use the SI system of units.

Now, consider Newton's Second Law

$$F = ma$$

It is an example of a general homogeneous physical equation since the dimensions on either side of the equality sign are $[LMT^{-2}]$. Its physical magnitudes can be expressed using any consistent system of units. Note that a general homogeneous physical equation does not contain a dimensional constant.[1(pp18−20)]

Consider an ancestor who described to his and her fellow cave mates how to make a spear. To demonstrate how long to make a spear, he or she placed a straight, trimmed sapling on the cave floor and ensured that its larger end touched the cave wall. Our ancestor then took his or her club and laid it beside the future spear, again ensuring that the end of the club touched the cave wall. Our ancestor then upended the club and walked it along the length of the future spear, counting each upending, until he or she reached its tip. Thus, our ancestor found the length of the future spear relative to the length of his or her club. Symbolically, our ancestor found

$$L_{Spear} = \alpha L_{Club}$$

where α is the number of times he or she upended the club from spear butt to tip. α is a pure number that we can manipulate with the logic and rules of mathematics. Note that L_{Spear} and L_{Club} are physical concepts; that is, they are symbols and are not subject to the logic and rules of mathematics. Looking at his or her fellow cave conferees, our ancestor realizes that clubs come in a variety of lengths. So, our ancestor decides to step-off the length of the future spear using his or her feet since most people have similar foot lengths. Our ancestor, therefore, backed against the cave wall and began stepping heel to toe along the length of the future spear, then he or she did the same along the length of his club. Our ancestor found that

$$L_{Spear} = \beta L_{foot}$$

and

$$L_{Club} = \gamma L_{foot}$$

Scratching his or her head, our ancestor realizes that the ratio of the future spear length to club length equals a pure number, namely

$$\frac{L_{\text{Spear}}}{L_{\text{Club}}} = \alpha$$

Our ancestor realizes the same is true for the second measurement, hence

$$\frac{L_{\text{Spear}}}{L_{\text{Club}}} = \frac{\beta \, L_{\text{foot}}}{\gamma \, L_{\text{foot}}}$$

But, the ratio of L_{foot} is constant and can be deleted from this ratio. Thus

$$\frac{L_{\text{Spear}}}{L_{\text{Club}}} = \frac{\beta}{\gamma}$$

Equating the two ratios, our ancestor obtained

$$\frac{L_{\text{Spear}}}{L_{\text{Club}}} = \frac{\beta}{\gamma} = \alpha$$

Since α, β, and γ are pure numbers, our ancestor realized that the ratio of two physical quantities, in this case L_{Spear} and L_{Club}, is equal to the ratio of the numbers of units used to measure them, regardless of the system of units used to measure them.[3] In other words, the ratio of physical magnitudes of similar dimension is independent of the system of units. Thus, the ratio of physical magnitudes possesses an absolute significance independent of the system of units used to measure the corresponding physical quantity.[2(p19)]

Note that the above result makes it inherent that physical magnitude is inversely proportional to the size of the unit used, which is due to the linearity of our fundamental dimensions.[4] This result brings us to the second axiom of Dimensional Analysis, which states

Axiom 2
*   The ratio of physical magnitudes of two like physical quantities $\alpha[\Psi]$ is independent of the system of units used to quantify them, so long as the numerator and denominator of the ratio use the same system of units.*[1(pp18−20)]

For example, let's assume our ancestor with the 50 ft by 100 ft garden plot has found a buyer for it. This buyer, unfortunately, lives in the neighboring kingdom where they measure length in "rods." The buyer has no idea what a foot length is and our ancestor has no idea what a rod length is. Therefore, the buyer brings his or her measuring rod to our ancestor's garden plot and finds it to be 3 rods by 6 rods.

The ratio of the length to breadth of our ancestor's garden plot, in the English Engineering system of units, is

$$\frac{100\ ft}{50\ ft} = 2$$

and in rods the ratio is

$$\frac{6\ rods}{3\ rods} = 2$$

as per Axiom 2. Note that the resulting ratios are dimensionless. Dividing one ratio by the other yields

$$\frac{100\ ft/50\ ft}{6\ rods/3\ rods} = \frac{2}{2} = 1$$

Thus, we can equate the two ratios, namely

$$\frac{100\ ft}{50\ ft} = \frac{6\ rods}{3\ rods}$$

which means that, within a given set of fundamental dimensions, all systems of units are equivalent. In other words, there is no distinguished or preferred system of units for a given set of fundamental dimensions.

We can also demonstrate Axiom 2 using a common engineering ratio. Consider the Reynolds number for fluid flow in a pipe, which is defined as

$$Re = \frac{\rho D v}{\mu}$$

where ρ is fluid density $[ML^{-3}]$; D is the pipe's diameter $[L]$; v is fluid velocity $[LT^{-1}]$; and μ is fluid dynamic viscosity $[L^{-1}MT^{-1}]$. In the English Engineering system of units, the density of water at 20°C is 62.3 lb_M/ft^3 and its viscosity is 2.36 lb_M/ft h or 0.000655 lb_M/ft s. If the pipe's diameter is 1 ft and the water is flowing at 100 ft/s, then the Reynolds number is

$$Re = \frac{(62.3\ lb_M/ft^3)(1\ ft)(100\ ft/s)}{0.000655\ lb_M/ft\ s} = \frac{6230\ lb_M/ft\ s}{0.000655\ lb_M/ft\ s} = 9.5 \times 10^6$$

In the SI system of units, water density at 20°C is 998 kg/m³ and its dynamic viscosity is 0.000977 kg/m s. The equivalent pipe diameter is 0.305 m and the equivalent water flow rate is 30.5 m/s. The Reynolds number is, then

$$Re = \frac{(998 \text{ kg/m}^3)(0.305 \text{ m})(30.5 \text{ m/s})}{0.000977 \text{ kg/m s}} = \frac{9284 \text{ kg/m s}}{0.000977 \text{ kg/m s}} = 9.5 \times 10^6$$

We can equate the above two ratios

$$\frac{(998 \text{kg/m}^3)(0.305\text{m})(30.5\text{m/s})}{0.000977\text{kg/ms}} = \frac{(62.3 \text{lb}_M/\text{ft}^3)(1\text{ft})(100\text{ft/s})}{0.000655 \text{lb}_M/\text{fts}} = 9.5 \times 10^6$$

which shows that the English Engineering system of units is equivalent to the SI system of units. This result again suggests that no distinguished or preferred system of units exists for the LMT set of dimensions.[5]

4.3 FOUNDATION OF METHOD OF INDICES

We can generalize this suggestion by considering a physical concept α that we want to quantify. Our first step is to choose a set of fundamental dimensions $[\Psi]$ that will quantify α. For example, let us choose Length, Mass, and Time (L, M, and T, respectively) as our fundamental dimension set. We next select the system of units we will use to determine the physical magnitude of α. Since there are many such systems of units, let us choose $L_1 M_1 T_1$ as our system of units. Thus

$$\alpha[LMT] = \Phi(L_1, M_1, T_1)$$

where α represents a physical concept and [LMT] represents the fundamental dimensions quantifying α. $\Phi(L_1, M_1, T_1)$ represents the function determining the physical magnitude in the chosen system of units. We could have chosen a different system of units, which we identify as $L_2 M_2 T_2$. Note that $L_1 M_1 T_1$ and $L_2 M_2 T_2$ are related by a constant, β, which we identify as a "conversion factor." Mathematically, the two systems of units are related as

$$\beta = \frac{\Phi(L_2, M_2, T_2)}{\Phi(L_1, M_1, T_1)}$$

Converting our physical quantity from the $L_1 M_1 T_1$ system of units to the $L_2 M_2 T_2$ system of units involves substituting $\Phi(L_2, M_2, T_2)/\beta$ for $\Phi(L_1, M_1, T_1)$, thus

$$\alpha[\text{LMT}] = \frac{\Phi(L_2, M_2, T_2)}{\beta}$$

Now, consider a third system of units designated $L_3 M_3 T_3$. Converting our physical quantity from the $L_1 M_1 T_1$ system of units to the $L_3 M_3 T_3$ system of units involves yet another conversion factor

$$\gamma = \frac{\Phi(L_3, M_3, T_3)}{\Phi(L_1, M_1, T_1)}$$

which upon substituting into

$$\alpha[\text{LMT}] = \Phi(L_1, M_1, T_1)$$

yields

$$\alpha[\text{LMT}] = \frac{\Phi(L_3, M_3, T_3)}{\gamma}$$

Dividing the last conversion by the previous conversion gives

$$\frac{\alpha[\text{LMT}]}{\alpha[\text{LMT}]} = \frac{\Phi(L_3, M_3, T_3)/\gamma}{\Phi(L_3, M_3, T_3)/\beta} = \frac{\beta\Phi(L_3, M_3, T_3)}{\gamma\Phi(L_2, M_2, T_2)} = 1$$

Thus

$$\frac{\gamma}{\beta} = \frac{\Phi(L_3, M_3, T_3)}{\Phi(L_2, M_2, T_2)}$$

Note that we could have done each of these conversions via a different route; namely, we could have converted each unit individually. Let us return to the conversion

$$\beta = \frac{\Phi(L_2, M_2, T_2)}{\Phi(L_1, M_1, T_1)}$$

and rearrange it. Doing so yields

$$\beta\Phi(L_1, M_1, T_1) = \Phi(L_2, M_2, T_2)$$

Dividing each term by its corresponding term in the first system of units, we get

$$\beta\Phi(L_1/L_1, M_1/M_1, T_1/T_1) = \Phi(L_2/L_1, M_2/M_1, T_2/T_1)$$

But

$$\Phi\left(L_1/L_1, M_1/M_1, T_1/T_1\right) = \Phi(1,1,1) = \kappa$$

where κ is a constant, which we define as 1. Thus

$$\beta = \Phi\left(L_2/L_1, M_2/M_1, T_2/T_1\right)$$

Similarly for the third system of units

$$\gamma = \Phi\left(L_3/L_1, M_3/M_1, T_3/T_1\right)$$

Dividing the above two conversions gives

$$\frac{\gamma}{\beta} = \frac{\Phi(L_3/L_1, M_3/M_1, T_3/T_1)}{\Phi(L_2/L_1, M_2/M_1, T_2/T_1)}$$

Multiplying each term by its corresponding ratio of first system of units to second system of units gives

$$\frac{\gamma}{\beta} = \frac{\Phi\left(\frac{L_3 L_1}{L_1 L_2}, \frac{M_3 M_1}{M_1 M_2}, \frac{T_3 T_1}{T_1 T_2}\right)}{\Phi\left(\frac{L_2 L_1}{L_1 L_2}, \frac{M_2 M_1}{M_1 M_2}, \frac{T_2 T_1}{T_1 T_2}\right)} = \frac{\Phi\left(\frac{L_3 L_1}{L_1 L_2}, \frac{M_3 M_1}{M_1 M_2}, \frac{T_3 T_1}{T_1 T_2}\right)}{1} = \Phi\left(\frac{L_3 L_1}{L_1 L_2}, \frac{M_3 M_1}{M_1 M_2}, \frac{T_3 T_1}{T_1 T_2}\right)$$

Simplifying the above equation yields

$$\frac{\gamma}{\beta} = \Phi(L_3/L_2, M_3/M_2, T_3/T_2)$$

Equating the two γ/β equations gives us

$$\frac{\Phi(L_3, M_3, T_3)}{\Phi(L_2, M_2, T_2)} = \Phi(L_3/L_2, M_3/M_2, T_3/T_2)$$

Differentiating the above equation with respect to L_3 gives

$$\frac{\partial\Phi(L_3, M_3, T_3)/\partial L_3}{\Phi(L_2, M_2, T_2)} = \frac{1}{L_2}\Phi(L_3/L_2, M_3/M_2, T_3/T_2)$$

When we let $L_1 = L_2 = L_3$, $M_1 = M_2 = M_3$, and $T_1 = T_2 = T_3$, the above equation becomes

$$\frac{d\Phi(L, M, T)/dL}{\Phi(L, M, T)} = \frac{1}{L}\Phi(1,1,1) = \frac{a}{L}$$

where $\Phi(111)$ is a constant designated as "a." Rearranging the above equation gives

$$\frac{d\Phi(L, M, T)}{\Phi(L, M, T)} = a\left(\frac{dL}{L}\right)$$

Integrating the above differential equation yields

$$\ln(\Phi(L, M, T)) = a \ln(L) + \ln(\Phi'(M, T))$$

or, in exponential notation

$$\Phi(L, M, T) = L^a \Phi'(M, T)$$

where $\Phi'(M,T)$ is a new function dependent upon M and T only. Performing the same operations on M and T eventually produces

$$\Phi(L, M, T) = \kappa L^a M^b T^c$$

But, κ is a constant equal to 1, therefore

$$\Phi(L, M, T) = L^a M^b T^c$$

Thus, the dimension function which determines the physical magnitude is a monomial power law, as purported by Lord Rayleigh in 1877.[2(chp 2),5,6]

4.4 DIMENSIONAL HOMOGENEITY

We will now use the above result to prove Fourier's comments about dimensional homogeneity. Consider a dependent variable y represented by a function of independent variables $x_1, x_2, x_3, \ldots, x_n$. This statement in mathematical notation is

$$y = f(x_1, x_2, x_3, \ldots, x_n)$$

Let us assume the function is the sum of its independent variables, thus

$$y = x_1 + x_2 + x_3 + \cdots + x_n$$

If the function represents a physical equation, then each term in the function has a dimension associated with it, namely

$$y[LMT] = x_1[L_1 M_1 T_1] + x_2[L_2 M_2 T_2] + x_3[L_3 M_3 T_3] + \cdots + x_n[L_n M_n T_n]$$

Substituting for y yields

$$(x_1 + x_2 + x_3 + \cdots + x_n)[LMT] = x_1[L_1M_1T_1] + x_2[L_2M_2T_2] + x_3[L_3M_3T_3] \\ + \cdots + x_n[L_nM_nT_n]$$

Expanding the terms to the left of the equality sign gives

$$x_1[LMT] + \cdots + x_n[LMT] = x_1[L_1M_1T_1] + \cdots + x_n[L_nM_nT_n]$$

Equating each term yields

$$x_1[LMT] = \quad x_1[L_1M_1T_1]$$
$$x_2[LMT] = \quad x_2[L_2M_2T_2]$$
$$\vdots \qquad\qquad \vdots$$
$$x_n[LMT] = \quad x_n[L_nM_nT_n]$$

But, from above

$$\alpha[\Psi] = \alpha[LMT] = \Phi(L, M, T) = L^aM^bT^c$$

Thus, the above set of linear equations becomes

$$x_1[LMT] = \quad L^aM^bT^c = \quad L^{a_1}M^{b_1}T^{c_1} = \quad x_1[L_1M_1T_1]$$
$$x_2[LMT] = \quad L^aM^bT^c = \quad L^{a_2}M^{b_2}T^{c_2} = \quad x_2[L_2M_2T_2]$$
$$\vdots \qquad\qquad \vdots \qquad\qquad \vdots \qquad\qquad \vdots$$
$$x_n[LMT] = \quad L^aM^bT^c = \quad L^{a_n}M^{b_n}T^{c_n} = \quad x_n[L_nM_nT_n]$$

Removing the leftmost and rightmost terms since they are superfluous yields

$$L^aM^bT^c = \quad L^{a_1}M^{b_1}T^{c_1}$$
$$L^aM^bT^c = \quad L^{a_2}M^{b_2}T^{c_2}$$
$$\vdots \qquad\qquad \vdots$$
$$L^aM^bT^c = \quad L^{a_n}M^{b_n}T^{c_n}$$

We can write the above set of equations more compactly as

$$L^aM^bT^c = L^{a_1}M^{b_1}T^{c_1} = \cdots = L^{a_n}M^{b_n}T^{c_n}$$

Equating like dimensions gives

$$L^a = L^{a_1} = \cdots = L^{a_n}$$
$$M^b = M^{b_1} = \cdots = M^{b_n}$$
$$T = T^{c_1} = T^{c_n}$$

which shows that, when adding, or subtracting, the dimension L, M, and T on each term must be the same.[7] In other words, we can only add apples to apples or oranges to oranges—we cannot add apples and oranges to get "orpels."

4.5 MATRIX FORMULATION OF DIMENSIONAL ANALYSIS

Consider a dependent variable y represented by a function of independent variables $x_1^{k_1}, x_2^{k_2}, \ldots, x_n^{k_n}$, where the ks are constants. Mathematically

$$y = f(x_1^{k_1}, x_2^{k_2}, \ldots, x_n^{k_n})$$

If we assume the function is the multiplicative product of the independent variables, then

$$y = x_1^{k_1} x_2^{k_2} \cdots x_n^{k_n}$$

If the function represents a physical equation, then

$$y[LMT] = x_1^{k_1}[L_1 M_1 T_1]^{k_1} x_2^{k_2}[L_2 M_2 T_2]^{k_2} \cdots x_n^{k_n}[L_n M_n T_n]^{k_n}$$

Substituting for y in the above equation gives

$$(x_1^{k_1} x_2^{k_2} \cdots x_n^{k_n})[LMT] = x_1^{k_1}[L_1 M_1 T_1]^{k_1} x_2^{k_2}[L_2 M_2 T_2]^{k_2} \cdots x_n^{k_n}[L_n M_n T_n]^{k_n}$$

Then dividing by $(x_1^{k_1} x_2^{k_2} \cdots x_n^{k_n})$ yields

$$[LMT] = \frac{x_1^{k_1}[L_1 M_1 T_1]^{k_1} x_2^{k_2}[L_2 M_2 T_2]^{k_2} \cdots x_n^{k_n}[L_n M_n T_n]^{k_n}}{(x_1^{k_1} x_2^{k_2} \cdots x_n^{k_n})}$$

$$= [L_1 M_1 T_1]^{k_1}[L_2 M_2 T_2]^{k_2} \cdots [L_n M_n T_n]^{k_n}$$

But, as previously stated

$$\alpha[\Psi] = \alpha[LMT] = \Phi(L, M, T) = L^a M^b T^c$$

and

$$\alpha[LMT]^{k_n} = \Phi(L, M, T)^{k_n} = (L^a M^b T^c)^{k_n}$$

Making this substitution into the third equation above yields

$$L^a M^b T^c = (L^{a_1} M^{b_1} T^{c_1})^{k_1} (L^{a_2} M^{b_2} T^{c_2})^{k_2} \cdots (L^{a_n} M^{b_n} T^{c_n})^{k_n}$$

Equating the exponential terms for L, M, and T, respectively, gives

$$a = a_1k_1 + a_2k_2 + \cdots + a_nk_n$$
$$b = b_1k_1 + b_2k_2 + \cdots + b_nk_n$$
$$c = c_1k_1 + c_2k_2 + \cdots + c_nk_n$$

Note that above, we have n terms but only three equations. Therefore, to solve this system of linear equations, we need to assume or assign values to n—3 terms. For convenience, let $n = 5$ in the above system of linear equations, then we have five unknowns and three equations; thus, we need to assume values for two unknowns. Let us assume we know k_3, k_4, and k_5. We will assume values for k_1 and k_2. We represent k_1 and k_2 as

$$k_1 = k_1 + 0 + 0 + \cdots + 0$$
$$k_2 = 0 + k_2 + 0 + \cdots + 0$$

Adding the k_1 and k_2 equations to the original set of linear equations gives us

$$k_1 = k_1 + 0 + 0 + \cdots + 0$$
$$k_2 = 0 + k_2 + 0 + \cdots + 0$$
$$a = a_1k_1 + a_2k_2 + \cdots + a_nk_n$$
$$b = b_1k_1 + b_2k_2 + \cdots + b_nk_n$$
$$c = c_1k_1 + c_2k_2 + \cdots + c_nk_n$$

which in matrix notation becomes

$$
\begin{bmatrix}
1 & 0 & 0 & 0 & 0 \\
0 & 1 & 0 & 0 & 0 \\
a_1 & a_2 & a_3 & a_4 & a_5 \\
b_1 & b_2 & b_3 & b_4 & b_5 \\
c_1 & c_2 & c_3 & c_4 & c_5
\end{bmatrix}
\begin{bmatrix}
k_1 \\ k_2 \\ k_3 \\ k_4 \\ k_5
\end{bmatrix}
=
\begin{bmatrix}
k_1 \\ k_2 \\ a \\ b \\ c
\end{bmatrix}
$$

From matrix algebra, we can partition the above matrices into

$$
\begin{bmatrix}
1 & 0 \\
0 & 1
\end{bmatrix}
$$

which is the Identity matrix or Unit matrix, represented by I, and the Zero matrix

$$
\begin{bmatrix}
0 & 0 & 0 \\
0 & 0 & 0
\end{bmatrix}
$$

represented by $0.^{8(\text{chp }7)}$ The matrix

$$\begin{bmatrix} a_1 & a_2 & a_3 & a_4 & a_5 \\ b_1 & b_2 & b_3 & b_4 & b_5 \\ c_1 & c_2 & c_3 & c_4 & c_5 \end{bmatrix}$$

is the Dimension matrix. It follows directly from the Dimension Table. The Dimension Table catalogs the dimensions of each variable of the original function. Thus, the Dimension Table has the below format

Variables		x_1	x_2	x_3	x_4	x_5
Dimensions	L	a_1	a_2	a_3	a_4	a_5
	M	b_1	b_2	b_3	b_4	b_5
	T	c_1	c_2	c_3	c_4	c_5

The Dimension matrix can be partitioned into two matrices, one being a square matrix, that is, a matrix with the same number of rows as columns; the other being the bulk or remaining matrix elements. We define the square matrix as the Rank matrix and the remaining matrix as the Bulk matrix. Partitioning the above Dimension matrix gives

$$\begin{bmatrix} \begin{bmatrix} a_1 & a_2 \\ b_1 & b_2 \\ c_1 & c_2 \end{bmatrix} & \begin{bmatrix} a_3 & a_4 & a_5 \\ b_3 & b_4 & b_5 \\ c_3 & c_4 & c_5 \end{bmatrix} \end{bmatrix}$$

We use the Rank matrix to calculate the "rank" of the Dimension matrix. We need the rank of the Dimension matrix in order to determine the number of independent solutions that exist for our system of linear equations. From linear algebra, the rank of a matrix is the number of linearly independent rows, or columns, of a matrix.[9] In other words, the rank of a matrix is the number of independent equations in a system of linear equations. Thus, the number of variables in a system of linear equations; that is, the number of columns in the Dimension matrix minus the rank of the Dimension matrix equals the number of selectable unknowns.[8(chp 7)] Mathematically

$$N_{\text{Var}} - R = N_{\text{Selectable}}$$

where R is the rank of the Dimension matrix.

To determine the Rank of the Dimension matrix, we must calculate the determinant of the Rank matrix. If the determinant of the Rank matrix is nonzero, then R is the number of rows or the number of columns in the

Rank matrix. The above Rank matrix is a 3×3 matrix; therefore, the rank of its Dimension matrix is 3. In this case, $N_{Var} = 5$ and $R = 3$; therefore $N_{Var} - R = N_{Selectable}$ is $5 - 3 = 2$. Therefore, to solve the above set of linear equations, we need to select two unknowns.

We can now rewrite the matrix equation

$$\begin{bmatrix} 1 & 0 & 0 & 0 & 0 \\ 0 & 1 & 0 & 0 & 0 \\ a_1 & a_2 & a_3 & a_4 & a_5 \\ b_1 & b_2 & b_3 & b_4 & b_5 \\ c_1 & c_2 & c_3 & c_4 & c_5 \end{bmatrix} \begin{bmatrix} k_1 \\ k_2 \\ k_3 \\ k_4 \\ k_5 \end{bmatrix} = \begin{bmatrix} k_1 \\ k_2 \\ a \\ b \\ c \end{bmatrix}$$

in terms of the partitioned matrices; the above matrix equation becomes

$$\begin{bmatrix} I & 0 \\ B & R \end{bmatrix} \begin{bmatrix} k_1 \\ k_2 \\ k_3 \\ k_4 \\ k_5 \end{bmatrix} = \begin{bmatrix} k_1 \\ k_2 \\ a \\ b \\ c \end{bmatrix}$$

Its solution is[8(chp 7)]

$$\begin{bmatrix} k_1 \\ k_2 \\ k_3 \\ k_4 \\ k_5 \end{bmatrix} = \begin{bmatrix} I & 0 \\ -R^{-1}B & R^{-1} \end{bmatrix} \begin{bmatrix} k_1 \\ k_2 \\ a \\ b \\ c \end{bmatrix}$$

The matrix just right of the equal sign is called the "Total" matrix. With regard to Dimensional Analysis, the number of columns in the Dimension matrix equals the number of variables in the system of linear equations and the difference between the number of columns in the Dimension matrix and the rank of the Dimension matrix equals the number of selectable unknowns in the system of linear equations. The number of selectable unknowns equals the number of columns in the Identity or Unit matrix I. The product of reading down a column of the Identity matrix is a dimensional or dimensionless parameter, depending upon our selection of a, b, and c. If we select $a = b = c = 0$, the parameters will be dimensionless. If a, b, and c are nonzero, then the parameters will have dimensions. For the former case,

$$N_{Var} - R = N_P$$

where N_P is the number of independent dimensionless parameters obtainable from a given set of linear equations. This result is known as Buckingham's Theorem or the Pi Theorem.[10-12] For the latter case,

$$N_{Var} - R + 1 = N_P$$

which is van Driest's rule, a variation of Buckingham's Theorem.[13]

4.6 IDENTIFYING VARIABLES FOR DIMENSIONAL ANALYSIS

The question always arises: how do we identify the variables for a Dimensional Analysis study? The best way to identify the variables for use in a Dimensional Analysis is to write the conservation laws and constitutive equations underpinning the process being studied. Constitutive equations describe a specific response of a given variable to an external force. The most familiar constitutive equations are Newton's Law of Viscosity, Fourier's Law of Heat Conduction, and Fick's Law of Diffusion.

The issue when identifying variables for a Dimensional Analysis study is not having too many, but missing pertinent ones. In the former situation, we still obtain the correct result; however, that result will contain extraneous variables, variables not actually required by Dimensional Analysis. In the latter situation, Dimensional Analysis produces an incorrect result. Therefore, to ensure the correct result, we will include any variable we deem remotely pertinent to the process being investigated. We can then identify, during our analysis of the process, which variables are irrelevant.

We determine which variables are irrelevant by calculating the matrix V, which is

$$V = (-R^{-1} \times B)^T$$

where superscript T identifies the "transpose" of the resulting matrix $-R^{-1} \times B$. A column of zeroes in matrix V identifies an irrelevant variable, a variable that can be dropped from the Dimensional Analysis of the process under study. We do not prove this assertion; it is proven elsewhere.[8(chp 8, 10, and 11)]

4.7 SUMMARY

As with all engineering and scientific endeavors, Dimensional Analysis involves procedure. Procedures are mechanisms that help us organize our thoughts. They are outlines of what we plan to do. As such, they minimize the likelihood that we will overlook or ignore an important point of our project. In other words, procedures reduce the time we expend on a given project and increase the accuracy of our result.

The procedure for using the matrix formulation of Dimensional Analysis includes

1. state the problem—clearly;
2. research all available literature for published results;
3. develop the pertinent balances, i.e., momentum, heat, and mass, for the problem;
4. list the important variables of the problem;
5. develop a Dimension Table using the identified variables;
6. write the Dimension matrix;
7. determine the Rank of the Dimension matrix;
8. identify the Rank matrix and calculate its inverse matrix;
9. identify the Bulk matrix;
10. multiply the negative of the inverse Rank matrix with the Bulk matrix;
11. determine whether irrelevant variables are in the Dimension matrix;
12. build the Total matrix;
13. read the dimensionless parameters from the Total matrix;
14. rearrange the dimensionless parameters to maximize physical content interpretation.

REFERENCES

1. G. Murphy, *Similitude in Engineering*, The Ronald Press Company, New York, NY, 1950.

2. P. Bridgman, *Dimensional Analysis*, Yale University Press, New Haven, CT, 1922.

3. J. Hunsaker, B. Rightmire, *Engineering Applications of Fluid Mechanics*, McGraw-Hill, New York, NY, 1947, Chapter 7.

4. H. Huntley, *Dimensional Analysis*, Dover Publications, Inc, New York, NY, 1967, page (first published by MacDonald and Company, Ltd., 1952).

5. G. Barenblatt, *Scaling*, Cambridge University Press, Cambridge, UK, 2003, pp. 17–20.

6. G. Barenblatt, *Scaling, Self-Similarity, and Intermediate Asymptotics*, Cambridge University Press, Cambridge, UK, 1996, pp. 34–37.

7. H. Langhaar, *Dimensional Analysis and Theory of Models*, John Wiley & Sons, New York, NY, 1951, Chapter 4.

8. T. Szirtes, *Applied Dimensional Analysis and Modeling*, Second Edition, Butterworth–Heinemann, Burlington, MA, 2007.

9. M. Jain, *Vector Spaces and Matrices in Physics*, Narosa Publishing House, New Delhi, India, 2001, p. 75.

10. E. Buckingham, *Physical Review*, *4* (4), 345 (1914).

11. L. Rayleigh, *Nature*, *95*, 66 (1915).

12. R. Tolman, *Physical Review*, *3*, 244 (1914).

13. E. Van Driest, *Journal of Applied Mechanics*, *13* (1), A-34 (1946).

CHAPTER 5

Mechanical/Physical Examples of Dimensional Analysis

5.1 INTRODUCTION

Aerodynamic, civil, and mechanical engineers have used Dimensional Analysis for gaining insight into complex, multivariable processes since the latter half of the nineteenth century. Aerodynamic and hydrodynamic problems involve the fundamental dimensions: Length [L], Mass [M], and Time [T], as do most mechanical engineering problems. Dimensional Analysis of aeronautical, civil, and mechanical problems can be done by "hand"; thus, the popularity and success of Dimensional Analysis in these engineering disciplines. The following examples are mechanical and physical in context. Hence, our fundamental dimensions will be Length, Mass, and Time. We will use the SI system of units when necessary. The following examples will also use exclusively the matrix formulation of Dimensional Analysis.

5.2 POINT MOVING IN A CIRCULAR ORBIT[1](PP48−49)

Consider a massless point moving in a circular orbit at uniform velocity. What is the functional relationship between acceleration and the other variables of this physical process? From experience and our knowledge of physics, we know this process has three variables, which are: acceleration, a [LT^{-2}]; velocity, v [LT^{-1}]; and orbit radius, r [L]. After identifying the variables, it is best to group them under general headings; e.g., geometric, material, and process variables. In this case, the geometric variable is r [L]. There are no material variables since the orbiting point is massless. The process variables are velocity v [LT^{-1}] and acceleration a [LT^{-2}]. The Dimension Table for this process is

Variables		a	v	r
Dimensions	L	1	1	1
	T	−2	−1	0

The Dimension matrix is

$$\begin{bmatrix} 1 & 1 & 1 \\ -2 & -1 & 0 \end{bmatrix}$$

We calculate the Rank of the Dimension matrix by ensuring that the furthest right portion of the Dimension matrix forms the largest possible square matrix possessing a nonsingular determinant. In this case, the largest square matrix is a 2×2 matrix. Next, we calculate the determinant of that square matrix. Since it is a 2×2 matrix, we can calculate its determinant by "hand"; the determinant is

$$\begin{vmatrix} 1 & 1 \\ -1 & 0 \end{vmatrix} = (1)(0) - (1)(-1) = 1$$

Thus, the determinant of the Rank matrix is nonsingular, i.e., it is not zero. Hence, the Rank of the Dimension matrix is 2 because its Rank matrix is a 2×2 matrix. Therefore, the number of dimensionless parameters for this process is

$$N_P = N_{Var} - R = 3 - 2 = 1$$

We must now calculate the inverse of the Rank matrix. Before the advent of the Internet, we made such calculations by "hand." Calculating the inverse of a 2×2 matrix can be done, but such a calculation becomes inordinately more difficult by "hand" as the square matrix grows. Much time is spent and many mistakes are made when inverting matrices by hand. However, today, we can simply go to the Internet and access a variety of free-for-use matrix calculators. Doing so gives us the inverse of the above Rank matrix, which is

$$R^{-1} = \begin{bmatrix} 1 & 1 \\ -1 & 0 \end{bmatrix}^{-1} = \begin{bmatrix} 0 & -1 \\ 1 & 1 \end{bmatrix}$$

For this process, the Bulk matrix is

$$B = \begin{bmatrix} 1 \\ -2 \end{bmatrix}$$

We can now calculate $-R^{-1} \times B$ using a matrix calculator on the Internet, which is

$$-R^{-1} \times B = -\begin{bmatrix} 0 & -1 \\ 1 & 1 \end{bmatrix}\begin{bmatrix} 1 \\ -2 \end{bmatrix} = \begin{bmatrix} -2 \\ 1 \end{bmatrix}$$

We next calculate matrix V to determine if our analysis includes any irrelevant variables. Matrix V is

$$V = (-R^{-1} \times B)^{\mathrm{T}} = \begin{bmatrix} -2 \\ 1 \end{bmatrix}^{\mathrm{T}} = [\,-2 \quad 1\,]$$

and, identifying the variables in matrix V

$$V = \begin{matrix} v & r \\ [\,-2 & 1\,] \end{matrix}$$

No column of matrix V is entirely comprised of zeroes; thus, we have no irrelevant variables in this example.

From the previous chapter, the matrix equation to solve for a massless point orbiting at uniform velocity is

$$\begin{bmatrix} k_1 \\ k_2 \\ k_3 \end{bmatrix} = \begin{bmatrix} I & 0 \\ -R^{-1} \times B & R^{-1} \end{bmatrix} \begin{bmatrix} k_1 \\ a \\ b \end{bmatrix}$$

But, for a dimensionless coefficient, $a = b = 0$. Thus, the above matrix equation becomes

$$\begin{bmatrix} k_1 \\ k_2 \\ k_3 \end{bmatrix} = \begin{bmatrix} I & 0 \\ -R^{-1} \times B & R^{-1} \end{bmatrix} \begin{bmatrix} k_1 \\ 0 \\ 0 \end{bmatrix}$$

Remember that k_1, k_2, and k_3 are the exponents or indices on acceleration a, velocity v, and radius r, respectively. It is best to inscribe them to the left of the k-matrix as a reminder; doing so, the above matrix equation becomes

$$\begin{matrix} a \\ v \\ r \end{matrix} \begin{bmatrix} k_1 \\ k_2 \\ k_3 \end{bmatrix} = \begin{bmatrix} I & 0 \\ -R^{-1} \times B & R^{-1} \end{bmatrix} \begin{bmatrix} k_1 \\ 0 \\ 0 \end{bmatrix}$$

The parameters a, v, and r in the above matrix equation are simply accounting indices; they do not enter the calculation. Completing the above matrix equation gives us

$$\begin{matrix} a \\ v \\ r \end{matrix} \begin{bmatrix} k_1 \\ k_2 \\ k_3 \end{bmatrix} = \begin{bmatrix} 1 & 0 & 0 \\ -2 & 0 & -1 \\ 1 & 1 & 1 \end{bmatrix} \begin{bmatrix} k_1 \\ 0 \\ 0 \end{bmatrix}$$

Since this example produces only one dimensionless parameter, the Unit matrix contains only one column. That column is the dimensionless parameter we are seeking and we identify it by placing Π_1 above it.

$$\Pi_1$$

$$
\begin{matrix} a \\ v \\ r \end{matrix}
\begin{bmatrix} k_1 \\ k_2 \\ k_3 \end{bmatrix}
=
\begin{bmatrix} 1 & 0 & 0 \\ -2 & 0 & -1 \\ 1 & 1 & 1 \end{bmatrix}
\begin{bmatrix} k_1 \\ 0 \\ 0 \end{bmatrix}
$$

To determine the resulting dimensionless parameter, all we have to do is read down the column of the Unit matrix. Thus, the dimensionless parameter Π_1 is

$$\Pi_1 = \frac{ar}{v^2}$$

The solution to this problem, in functional notation, is

$$f(\Pi_1) = f\left(\frac{ar}{v^2}\right) = 0$$

or, in terms of acceleration, we get

$$\frac{ar}{v^2} = \kappa$$

which, upon rearrangement, is

$$a = \kappa\left(\frac{v^2}{r}\right)$$

which shows that the acceleration of the massless point along its circular orbit depends upon v^2/r and a constant κ, both of which are determined by experiment.

5.3 VOLUMETRIC FLOW RATE OF A FLUID THROUGH A TUBE[1(PP54−55)]

Consider a viscous, incompressible fluid flowing through a tube of circular cross section. Liquid flow through the tube occurs due to a pressure difference gradient between the two ends of the tube. What is the relationship of volumetric flow rate Q [L^3T^{-1}] to the other variables of the process? The geometric variable is tube length l [L] and tube radius r [L]. The material variable is fluid viscosity μ [$L^{-1}MT^{-1}$]. The process variable is pressure gradient $(p_{Up} - p_{Down})/l$ [$L^{-2}MT^{-2}$], where p_{Up} is the upstream pressure and p_{Down}, the downstream pressure of the flowing fluid.

The Dimension Table is

Variables		Q	$\Delta p/l$	r	μ
Dimensions	L	3	−2	1	−1
	M	0	1	0	1
	T	−1	−2	0	−1

We can immediately write the Dimension matrix, which is

$$\begin{bmatrix} 3 & -2 & 1 & -1 \\ 0 & 1 & 0 & 1 \\ -1 & -2 & 0 & -1 \end{bmatrix}$$

and, starting from the furthest right column of the Dimension matrix, we write the largest possible square matrix; it is a 3×3 matrix, which is our Rank matrix

$$R = \begin{bmatrix} -2 & 1 & -1 \\ 1 & 0 & 1 \\ -2 & 0 & -1 \end{bmatrix}$$

We could calculate the determinant of R by hand, but it is easier and less error prone to use one of the free-for-use Internet matrix calculators available to us. Doing so yields

$$|R| = \begin{vmatrix} -2 & 1 & -1 \\ 1 & 0 & 1 \\ -2 & 0 & -1 \end{vmatrix} = -1$$

Thus, the Rank of the Dimension matrix is 3 since the Rank matrix is a 3×3 matrix. Therefore, the number of dimensionless parameters will be

$$N_P = N_{Var} - R = 4 - 3 = 1$$

The Bulk matrix is

$$B = \begin{bmatrix} 3 \\ 0 \\ -1 \end{bmatrix}$$

The inverse of the Rank matrix is

$$R^{-1} = \begin{bmatrix} -2 & 1 & -1 \\ 1 & 0 & 1 \\ -2 & 0 & -1 \end{bmatrix}^{-1} = \begin{bmatrix} 0 & -1 & -1 \\ 1 & 0 & -1 \\ 0 & 2 & 1 \end{bmatrix}$$

and the product of $-R^{-1} \times B$ is

$$-R^{-1} \times B = - \begin{bmatrix} 0 & -1 & -1 \\ 1 & 0 & -1 \\ 0 & 2 & 1 \end{bmatrix} \begin{bmatrix} 3 \\ 0 \\ -1 \end{bmatrix} = \begin{bmatrix} -1 \\ -4 \\ 1 \end{bmatrix}$$

We next calculate matrix V to determine if our analysis includes any irrelevant variables. Matrix V is

$$V = (-R^{-1} \times B)^{\mathrm{T}} = \begin{bmatrix} -1 \\ -4 \\ 1 \end{bmatrix}^{\mathrm{T}} = [-1 \quad -4 \quad 1]$$

and, identifying the variables in matrix V

$$\begin{array}{ccc} \Delta p/l & r & \mu \\ V = [-1 & -4 & 1] \end{array}$$

No column of matrix V is entirely comprised of zeroes; thus, we have no irrelevant variables in this example.

We now have all the partition matrices necessary to build the matrix equation to be solved. That equation is

$$\begin{array}{c} Q \\ \Delta p/l \\ r \\ \mu \end{array} \begin{bmatrix} k_1 \\ k_2 \\ k_3 \\ k_4 \end{bmatrix} = \begin{bmatrix} I & 0 \\ -R^{-1} \times B & R^{-1} \end{bmatrix} \begin{bmatrix} k_1 \\ a \\ b \\ c \end{bmatrix}$$

where $k_1, \ldots, k_4$ are the exponents or indices on the variables which are listed to the far left of the matrix equation. The listed variables do not enter the calculation. We can now complete the entire matrix equation, which is

$$\begin{array}{c} \\ Q \\ \Delta p/l \\ r \\ \mu \end{array} \begin{array}{c} \Pi_1 \\ \begin{bmatrix} k_1 \\ k_2 \\ k_3 \\ k_4 \end{bmatrix} = \begin{bmatrix} 1 & 0 & 0 & 0 \\ -1 & 0 & -1 & -1 \\ -4 & 1 & 0 & -1 \\ 1 & 0 & 2 & 0 \end{bmatrix} \begin{bmatrix} k_1 \\ 0 \\ 0 \\ 0 \end{bmatrix} \end{array}$$

where $a = b = c = 0$ since we desire a dimensionless result. Since this example produces only one dimensionless parameter, the Unit matrix contains only one column. We identify that column in the above matrix equation by placing Π_1 above it. Reading down the Unit matrix column yields the dimensionless parameter, which is

$$\Pi_1 = \frac{Q\mu}{(\Delta p/l)r^4}$$

In functional notation, the solution to this example is

$$f(\Pi_1) = f\left(\frac{Q\mu}{(\Delta p/l)r^4}\right) = 0$$

We can write the above function as

$$\frac{Q\mu}{(\Delta p/l)r^4} = \kappa$$

Solving for Q gives

$$Q = \kappa \times \frac{(\Delta p/l)r^4}{\mu}$$

κ is found by experiment to be $\pi/8$. The final equation is

$$Q = \frac{\pi(\Delta p/l)r^4}{8\mu}$$

which is Poiseuille's law for fluid flow through a circular tube.

5.4 VELOCITY ON AN INCLINED PLANE[2](PP299–301)

Consider the proverbial massless sphere on an inclined plane. The massless sphere rests on the plane at h, which is an unstable condition. With the slightest perturbation, the sphere will begin moving down the incline. The geometric variables are sphere radius r [L] and the height h [L] of the inclined plane above the horizontal datum plane. There are no material variables since the sphere is massless. The process variables are the terminal velocity v_T [LT^{-1}] that the sphere acquires upon rolling down the incline and gravitational acceleration g [LT^{-2}]. What is the relationship between these variables?

The Dimension Table is

Variables		v_T	r	g	h
Dimensions	L	1	1	1	1
	T	−1	0	−2	0

The Dimension matrix is

$$\begin{bmatrix} 1 & 1 & 1 & 1 \\ -1 & 0 & -2 & 0 \end{bmatrix}$$

Starting at the far left of the Dimension matrix and moving rightward, the largest square matrix is a 2×2 matrix; thus, the Rank matrix is

$$R = \begin{bmatrix} 1 & 1 \\ -2 & 0 \end{bmatrix}$$

The determinant of the Rank matrix is

$$|R| = \begin{vmatrix} 1 & 1 \\ -2 & 0 \end{vmatrix} = -2$$

A nonsingular determinant means the Rank of the Dimension matrix is 2. Therefore, the number of dimensionless parameters will be

$$N_P = N_{Var} - R = 4 - 2 = 2$$

The Bulk matrix is

$$B = \begin{bmatrix} 1 & 1 \\ -1 & 0 \end{bmatrix}$$

The inverse of the Rank matrix is

$$R^{-1} = \begin{bmatrix} 1 & 1 \\ -2 & 0 \end{bmatrix} = \begin{bmatrix} 0 & -0.5 \\ 1 & 0.5 \end{bmatrix}$$

and the product of $-R^{-1} \times B$ is

$$-R^{-1} \times B = -\begin{bmatrix} 0 & -0.5 \\ 1 & 0.5 \end{bmatrix}\begin{bmatrix} 1 & 1 \\ -1 & 0 \end{bmatrix} = \begin{bmatrix} -0.5 & 0 \\ -0.5 & -1 \end{bmatrix}$$

We next calculate matrix V to determine if our analysis includes any irrelevant variables. Matrix V is, with variable identification,

$$V = (-R^{-1} \times B)^T = \begin{bmatrix} -0.5 & 0 \\ -0.5 & -1 \end{bmatrix}^T = \overset{\begin{matrix} g & h \end{matrix}}{\begin{bmatrix} -0.5 & -0.5 \\ 0 & -1 \end{bmatrix}}$$

No column of matrix V is entirely comprised of zeroes; thus, we have no irrelevant variables in this example.

We now have all the partition matrices necessary to build the matrix equation to be solved. That equation is

$$\begin{matrix} v_T \\ r \\ g \\ h \end{matrix} \begin{bmatrix} k_1 \\ k_2 \\ k_3 \\ k_4 \end{bmatrix} = \begin{bmatrix} I & 0 \\ -R^{-1} \times B & R^{-1} \end{bmatrix} \begin{bmatrix} k_1 \\ a \\ b \\ c \end{bmatrix}$$

where $k_1, \ldots, k_4$ are the exponents or indices on the variables which are listed to the far left of the matrix equation. The listed variables do not enter into the calculation. We can now complete the entire matrix equation, which is

$$\begin{matrix} & & \Pi_1 & \Pi_2 \\ v_T \\ r \\ g \\ h \end{matrix} \begin{bmatrix} k_1 \\ k_2 \\ k_3 \\ k_4 \end{bmatrix} = \begin{bmatrix} 1 & 0 & 0 & 0 \\ 0 & 1 & 0 & 0 \\ -0.5 & 0 & 0 & -0.5 \\ -0.5 & -1 & 1 & 0.5 \end{bmatrix} \begin{bmatrix} k_1 \\ k_2 \\ 0 \\ 0 \end{bmatrix}$$

Since the result will contain two dimensionless parameters, the Unit matrix will have two columns. Those columns are the two rightmost columns of the Total matrix. We identify those columns by placing Π_1 and Π_2 above them. Reading down those columns and matching exponent with the appropriate variable identifies the two dimensionless parameters, which are

$$\Pi_1 = \frac{v_T}{\sqrt{gh}} \text{ and } \Pi_2 = \frac{r}{h}$$

The question now becomes: are these two dimensionless parameters independent of each other? From linear algebra, the test for independence is

$$\alpha \Pi_1 + \beta \Pi_2 = 0$$

Substituting the matrix Π_1 and Π_2 columns into the above equation gives

$$\alpha \begin{bmatrix} 1 \\ 0 \\ -0.5 \\ -0.5 \end{bmatrix} + \beta \begin{bmatrix} 0 \\ 1 \\ 0 \\ -1 \end{bmatrix} = \begin{bmatrix} 0 \\ 0 \\ 0 \\ 0 \end{bmatrix}$$

For Π_1 and Π_2 to be independent, α and β must both be zero. Solving the above matrix equation yields

$$\alpha + 0 = 0$$
$$0 + \beta = 0$$
$$-0.5\alpha + 0 = 0$$
$$-0.5\alpha - \beta = 0$$

The above system of linear equations shows that $\alpha = \beta = 0$; thus, Π_1 and Π_2 are independent of each other.

The solution to this example, in functional notation, is

$$f(\Pi_1, \Pi_2) = f\left(\frac{v_T}{\sqrt{gh}}, \frac{r}{h}\right) = 0$$

or, we can distinguish the dimensionless coefficient, which contains the dependent variable, in functional notation as

$$\Pi_1 = f(\Pi_2)$$

If we know one variable is linearly dependent upon another variable, then we need to perform only three experiments to determine the slope and intercept of the function. In general, however, we do not know whether a given variable depends linearly on another variable or not. In such cases, we must perform at least five experiments to determine the curvature of the function or to determine whether the function is oscillatory. Thus, for multivariable functions, the number of experiments to define the function grows rapidly because we must perform five experiments per independent variable while maintaining all other variables constant. Therefore, the total number of experiments $N_{\text{Expts}}^{\text{Total}}$ required to define a function is

$$N_{\text{Expts}}^{\text{Total}} = N_{\text{Expts}}^{M_{\text{IndepVar}}}$$

where N_{Expts} is the number of experiments to be performed per independent variable and M_{IndepVar} is the number of independent variables in the function.

For this example, the dimensionless function is

$$\Pi_1 = f(\Pi_2)$$

Thus, we have one dependent variable and one independent variable. The total number of experiments required to define this function is

$$N_{\text{Expts}}^{\text{Total}} = N_{\text{Expts}}^{M_{\text{IndepVar}}} = 5^1 = 5$$

If we solve this example dimensionally, the solution function is

$$v_T = f(r, g, h)$$

Thus, $N_{\text{Expts}}^{\text{Total}}$ is

$$N_{\text{Expts}}^{\text{Total}} = N_{\text{Expts}}^{M_{\text{IndepVar}}} = 5^2 = 25$$

g is a constant; therefore, it does not generate any experiments. In other words, we perform 20 fewer experiments by solving this example using dimensionless parameters compared to solving it dimensionally.

5.5 WATER FLOW OVER A WEIR[2(PP335–336)]

Many chemical plants have storm water runoff reservoirs that allow material washed from the site to settle, thereby sequestering such material from the environment. A weir maintains these ponds at a given water level. When the sequestered storm water reaches the top of the weir, it begins to flow over it. Our goal is to determine the mass flow rate of water spilling over the weir. The geometric variables are weir width w [L] and water height h [L] flowing over the weir. Together, w and h determine the cross sectional area of the water flowing over the weir. The material variable is water density ρ [L^{-3}M]. The process variables are mass flow rate Q [MT^{-1}] and gravitational acceleration g [LT^{-2}].

The Dimension Table is

Variables		Q	w	h	g	ρ
Dimensions	L	0	1	1	1	-3
	M	1	0	0	0	1
	T	-1	0	0	-2	0

from which the Dimension matrix is

$$\begin{bmatrix} 0 & 1 & 1 & 1 & -3 \\ 1 & 0 & 0 & 0 & 1 \\ -1 & 0 & 0 & -2 & 0 \end{bmatrix}$$

The Rank matrix is the largest square matrix, with a non-singular determinant, contained in the Dimension matrix. Starting with the rightmost column of the above Dimension matrix, the Rank matrix is

$$R = \begin{bmatrix} 1 & 1 & -3 \\ 0 & 0 & 1 \\ 0 & -2 & 0 \end{bmatrix}$$

and its determinant, as calculated by a free-for-use matrix calculator on the Internet, is

$$|R| = \begin{vmatrix} 1 & 1 & -3 \\ 0 & 0 & 1 \\ 0 & -2 & 0 \end{vmatrix} = 2$$

which is nonsingular. Therefore, the Rank of the above Dimension matrix is 3. The number of dimensionless parameters generated for this example will be

$$N_P = N_{Var} - R = 5 - 3 = 2$$

The inverse of the Rank matrix is

$$R^{-1} = \begin{bmatrix} 1 & 1 & -3 \\ 0 & 0 & 1 \\ 0 & -2 & 0 \end{bmatrix}^{-1} = \begin{bmatrix} 1 & 3 & 0.5 \\ 0 & 0 & -0.5 \\ 0 & 1 & 0 \end{bmatrix}$$

Note that the Bulk matrix is

$$B = \begin{bmatrix} 0 & 1 \\ 1 & 0 \\ -1 & 0 \end{bmatrix}$$

The product of $-R^{-1} \times B$ is

$$-R^{-1} \times B = -\begin{bmatrix} 1 & 3 & 0.5 \\ 0 & 0 & -0.5 \\ 0 & 1 & 0 \end{bmatrix} \begin{bmatrix} 0 & 1 \\ 1 & 0 \\ -1 & 0 \end{bmatrix} = \begin{bmatrix} -2.5 & -1 \\ -0.5 & 0 \\ -1 & 0 \end{bmatrix}$$

Calculating matrix V to determine if our analysis includes any irrelevant variables, we obtain, with variable identification,

$$V = (-R^{-1} \times B)^T = \begin{bmatrix} -2.5 & -1 \\ -0.5 & 0 \\ -1 & 0 \end{bmatrix}^T = \overset{\begin{matrix} h & g & \rho \end{matrix}}{\begin{bmatrix} -2.5 & -0.5 & -1 \\ -1 & 0 & 0 \end{bmatrix}}$$

No column of matrix V is entirely comprised of zeroes; thus, we have no irrelevant variables in this example.

We can now assemble the matrix equation and the Total matrix for this example. The matrix equation is

$$
\begin{bmatrix} k_1 \\ k_2 \\ k_3 \\ k_4 \\ k_5 \end{bmatrix} = \begin{bmatrix} I & 0 \\ -R^{-1} \times B & R^{-1} \end{bmatrix} \begin{bmatrix} k_1 \\ k_2 \\ a \\ b \\ c \end{bmatrix}
$$

and the Total matrix is

$$
T = \begin{bmatrix} I & 0 \\ -R^{-1} \times B & R^{-1} \end{bmatrix} = \begin{bmatrix} 1 & 0 & 0 & 0 & 0 \\ 0 & 1 & 0 & 0 & 0 \\ -2.5 & -1 & 1 & 3 & 0.5 \\ -0.5 & 0 & 0 & 0 & -0.5 \\ -1 & 0 & 0 & 1 & 0 \end{bmatrix}
$$

Since we read the dimensionless parameters from the Total matrix, we will cease writing the entire matrix equation and present only the Total matrix. This example produces two dimensionless parameters; thus, the Unit matrix for it contains only two columns, which we identify by placing Π_1 and Π_2 above the appropriate matrix columns.

$$
\begin{array}{c}
 \quad \Pi_1 \quad \Pi_2 \\
T = \begin{array}{c} Q \\ w \\ h \\ g \\ \rho \end{array} \begin{bmatrix} 1 & 0 & 0 & 0 & 0 \\ 0 & 1 & 0 & 0 & 0 \\ -2.5 & -1 & 1 & 3 & 0.5 \\ -0.5 & 0 & 0 & 0 & -0.5 \\ -1 & 0 & 0 & 1 & 0 \end{bmatrix}
\end{array}
$$

We have listed the variables for this example to the left of the Total matrix for accounting purposes. Reading down each identified column provides us with the dimensionless parameters, namely

$$
\Pi_1 = \frac{Q}{\rho h^{2.5} \sqrt{g}} = \frac{Q}{\rho \sqrt{g h^5}} \quad \text{and} \quad \Pi_2 = \frac{w}{h}
$$

The question now becomes: are these two dimensionless parameters independent of each other? From linear algebra, the test for independence is

$$
\alpha \Pi_1 + \beta \Pi_2 = 0
$$

Substituting the matrix Π_1 and Π_2 columns into the above equation gives

$$\alpha \begin{bmatrix} 1 \\ 0 \\ -2.5 \\ -0.5 \\ -1 \end{bmatrix} + \beta \begin{bmatrix} 0 \\ 1 \\ -1 \\ 0 \\ 0 \end{bmatrix} = \begin{bmatrix} 0 \\ 0 \\ 0 \\ 0 \\ 0 \end{bmatrix}$$

For Π_1 and Π_2 to be independent, α and β must both be zero. Solving the above matrix equation yields

$$\alpha + 0 = 0$$
$$0 + \beta = 0$$
$$-2.5\alpha - \beta = 0$$
$$-0.5\alpha + 0 = 0$$
$$-\alpha + 0 = 0$$

The above system of linear equations shows that $\alpha = \beta = 0$; thus, Π_1 and Π_2 are independent of each other.

In functional notation, the solution is

$$f(\Pi_1, \Pi_2) = f\left(\frac{Q}{\rho\sqrt{gh^5}}, \frac{w}{h} \right) = 0$$

or, solving for Q, the functional solution becomes

$$\Pi_1 = \kappa \times f(\Pi_2)$$

$$\frac{Q}{\rho\sqrt{gh^5}} = \kappa \times f\left(\frac{w}{h} \right)$$

$$Q = \kappa \times \rho\sqrt{gh^5} \times f\left(\frac{w}{h} \right)$$

where κ and $f(w/h)$ are determined by experimentation.

5.6 DETERMINING FLUID VISCOSITY

Viscosity is the constant of proportionality relating shear stress and shear rate deformation. Viscosity can be considered momentum conductivity. In this example, we determine the viscosity of a fluid by measuring the descent velocity of metal spheres through the fluid.

The geometric variable is metal sphere diameter D [L]. The material variable is fluid viscosity μ [$L^{-1}MT^{-1}$]. The process variables are: difference in metal sphere density and fluid density $\Delta\rho = \rho_S - \rho_F$, where ρ_S is the density of the metal sphere and ρ_F is the density of the fluid; gravitational acceleration g [LT^{-2}]; and terminal velocity of the metal sphere v [LT^{-1}].

The Dimension Table is

Variables		μ	D	v	$\Delta\rho$	g
Dimensions	L	−1	1	1	−3	1
	M	1	0	0	1	0
	T	−1	0	−1	0	−2

The Dimension matrix is

$$\begin{bmatrix} -1 & 1 & 1 & -3 & 1 \\ 1 & 0 & 0 & 1 & 0 \\ -1 & 0 & -1 & 0 & -2 \end{bmatrix}$$

The Rank matrix is the largest, nonsingular, square matrix contained in the Dimension matrix. The Rank matrix should be placed in the rightmost section of the Dimension matrix. The Rank matrix for this example is

$$R = \begin{bmatrix} 1 & -3 & 1 \\ 0 & 1 & 0 \\ -1 & 0 & -2 \end{bmatrix}$$

Its determinant, using a free-for-use matrix calculator available on the Internet, is

$$|R| = \begin{vmatrix} 1 & -3 & 1 \\ 0 & 1 & 0 \\ -1 & 0 & -2 \end{vmatrix} = -1$$

Therefore, the Rank of the Dimension matrix is 3, which means this example involves

$$N_P = N_{Var} - R = 5 - 3 = 2$$

dimensionless parameters. The inverse of the Rank matrix is

$$R^{-1} = \begin{bmatrix} 1 & -3 & 1 \\ 0 & 1 & 0 \\ -1 & 0 & -2 \end{bmatrix}^{-1} = \begin{bmatrix} 2 & 6 & 1 \\ 0 & 1 & 0 \\ -1 & -3 & -1 \end{bmatrix}$$

The Bulk matrix is

$$B = \begin{bmatrix} -1 & 1 \\ 1 & 0 \\ -1 & 0 \end{bmatrix}$$

and $-R^{-1} \times B$ is

$$-R^{-1} * B = -\begin{bmatrix} 2 & 6 & 1 \\ 0 & 1 & 0 \\ -1 & -3 & -1 \end{bmatrix}\begin{bmatrix} -1 & 1 \\ 1 & 0 \\ -1 & 0 \end{bmatrix} = \begin{bmatrix} -3 & -2 \\ -1 & 0 \\ 1 & 1 \end{bmatrix}$$

Calculating matrix V to determine if our analysis includes any irrelevant variables, we obtain, with variable identification,

$$V = (-R^{-1} \times B)^{\mathrm{T}} = \begin{bmatrix} -2.5 & -1 \\ -0.5 & 0 \\ -1 & 0 \end{bmatrix}^{\mathrm{T}} = \begin{matrix} v & \Delta\rho & g \\ \begin{bmatrix} -3 & -1 & 1 \\ -2 & 0 & 1 \end{bmatrix} \end{matrix}$$

No column of matrix V is entirely comprised of zeroes; thus, we have no irrelevant variables in this example.

The Total matrix for this example is

$$T = \begin{matrix} & \Pi_1 & \Pi_2 & & & \\ \mu & \\ D & \\ v & \\ \Delta\rho & \\ g & \end{matrix}\begin{bmatrix} 1 & 0 & 0 & 0 & 0 \\ 0 & 1 & 0 & 0 & 0 \\ -3 & -2 & 2 & 6 & 1 \\ -1 & 0 & 0 & 1 & 0 \\ 1 & 1 & -1 & -3 & -1 \end{bmatrix}$$

Note that we have listed the variables along the left side of the Total matrix. They do not enter into any calculation—they are presented for accounting reasons.

The dimensionless parameters are

$$\Pi_1 = \frac{\mu g}{v^3 \Delta\rho} \quad \text{and} \quad \Pi_2 = \frac{gD}{v^2}$$

The question now becomes: are these two dimensionless parameters independent of each other? The test for linear independence is

$$\alpha \Pi_1 + \beta \Pi_2 = 0$$

Substituting the matrix Π_1 and Π_2 columns into the above equation gives

$$\alpha \begin{bmatrix} 1 \\ 0 \\ -3 \\ -1 \\ 1 \end{bmatrix} + \beta \begin{bmatrix} 0 \\ 1 \\ -2 \\ 0 \\ 1 \end{bmatrix} = \begin{bmatrix} 0 \\ 0 \\ 0 \\ 0 \\ 0 \end{bmatrix}$$

For Π_1 and Π_2 to be independent, α and β must both be zero. Solving the above matrix equation yields

$$\alpha + 0 = 0$$
$$0 + \beta = 0$$
$$-3\alpha - 2\beta = 0$$
$$-\alpha + 0 = 0$$
$$\alpha + \beta = 0$$

The above system of linear equations shows that $\alpha = \beta = 0$; thus, Π_1 and Π_2 are independent of each other.

In functional notation, the solution is

$$f(\Pi_1, \Pi_2) = 0 \quad \text{or} \quad \Pi_1 = \kappa \times f(\Pi_2)$$

The functional solution, in terms of viscosity, is

$$\frac{\mu g}{v^3 \Delta \rho} = \kappa \times f(\Pi_2)$$

which upon rearrangement gives

$$\mu = \kappa \times \left(\frac{v^3 \Delta \rho}{g} \right) \times f(\Pi_2)$$

where κ and $f(\Pi_2)$ are determined by experimentation.

5.7 KEPLER'S SECOND LAW[2(PP360−363)]

Kepler's Second Law states that the vector from the Sun to an orbiting plant, such as the Earth, sweeps equal areas in equal time. We will use Dimensional Analysis to demonstrate this physical law.

The geometric variables are the area swept by the vector per unit time A $[L^2T^{-1}]$ and length of the Sun-to-Earth vector $L_{SunEarth}$ $[L]$. The material variable is the Sun's mass M_{Sun} $[M]$. The process variables are: time t $[T]$, determined from the orbit perihelion, which is the point of Earth's orbit nearest to the Sun's center; Earth's speed at the perihelion v $[LT^{-1}]$; and the universal gravitational constant G $[L^3M^{-1}T^{-2}]$.

The Dimension Table is

Variables		t	A	v	$L_{SunEarth}$	M_{Sun}	G
Dimensions	L	0	2	1	1	0	3
	M	0	0	0	0	1	−1
	T	1	−1	−1	0	0	−2

and the Dimension matrix is

$$\begin{bmatrix} 0 & 2 & 1 & 1 & 0 & 3 \\ 0 & 0 & 0 & 0 & 1 & -1 \\ 1 & -1 & -1 & 0 & 0 & -2 \end{bmatrix}$$

The Rank matrix is the largest, nonsingular, square matrix contained in the Dimension matrix. The Rank matrix for this example is

$$R = \begin{bmatrix} 1 & 0 & 3 \\ 0 & 1 & -1 \\ 0 & 0 & 2 \end{bmatrix}$$

Its determinant, using a free-for-use matrix calculator available on the Internet, is

$$|R| = \begin{vmatrix} 1 & 0 & 3 \\ 0 & 1 & -1 \\ 0 & 0 & 2 \end{vmatrix} = -2$$

Therefore, the Rank of the Dimension matrix is 3, which means this demonstration involves

$$N_P = N_{Var} - R = 6 - 3 = 3$$

dimensionless parameters. The inverse of the Rank matrix is

$$R^{-1} = \begin{bmatrix} 1 & 0 & 3 \\ 0 & 1 & -1 \\ 0 & 0 & 2 \end{bmatrix}^{-1} = \begin{bmatrix} 1 & 0 & 1.5 \\ 0 & 1 & -0.5 \\ 0 & 0 & -0.5 \end{bmatrix}$$

Our Bulk matrix is

$$B = \begin{bmatrix} 0 & 2 & 1 \\ 0 & 0 & 0 \\ 1 & -1 & -1 \end{bmatrix}$$

and $-R^{-1} \times B$ is

$$-R^{-1} \times B = \begin{bmatrix} -1 & 0 & -1.5 \\ 0 & -1 & 0.5 \\ 0 & 0 & 0.5 \end{bmatrix} \begin{bmatrix} 0 & 2 & 1 \\ 0 & 0 & 0 \\ 1 & -1 & -1 \end{bmatrix} = \begin{bmatrix} -1.5 & -0.5 & 0.5 \\ 0.5 & -0.5 & -0.5 \\ 0.5 & -0.5 & -0.5 \end{bmatrix}$$

Since there are no zeroes in $-R^{-1} \times B$, there will be no zeroes in $(-R^{-1} \times B)^{\mathrm{T}}$. Thus, there are no irrelevant variables in this example.

The Total matrix for this example is

$$
T = \begin{array}{c} \\ t \\ A \\ v \\ L_{\text{SunEarth}} \\ M_{\text{Sun}} \\ G \end{array}
\begin{array}{c} \Pi_1 \quad \Pi_2 \quad \Pi_3 \\ \begin{bmatrix} 1 & 0 & 0 & 0 & 0 & 0 \\ 0 & 1 & 0 & 0 & 0 & 0 \\ 0 & 0 & 1 & 0 & 0 & 0 \\ -1.5 & -0.5 & 0.5 & 1 & 0 & 1.5 \\ 0.5 & -0.5 & -0.5 & 0 & 1 & -0.5 \\ 0.5 & -0.5 & -0.5 & 0 & 0 & -0.5 \end{bmatrix} \end{array}
$$

The first three rows of those columns contain the Unit matrix, which we identify with Π_1, Π_2, and Π_3. The three dimensionless parameters are

$$\Pi_1 = \frac{t\sqrt{GM_{\text{Sun}}}}{L_{\text{SunEarth}}^{1.5}} = t\sqrt{\frac{GM_{\text{Sun}}}{L_{\text{SunEarth}}^3}}; \quad \Pi_2 = \frac{A}{\sqrt{GL_{\text{SunEarth}}M_{\text{Sun}}}}; \quad \text{and}$$

$$\Pi_3 = \frac{v\sqrt{L_{\text{SunEarth}}}}{\sqrt{GM_{\text{Sun}}}} = v\sqrt{\frac{L_{\text{SunEarth}}}{GM_{\text{Sun}}}}$$

So, are these three dimensionless parameters independent of each other? The test for linear independence is

$$\alpha\Pi_1 + \beta\Pi_2 + \gamma\Pi_3 = 0$$

Substituting the matrix Π_1, Π_2, and Π_3 columns into the above equation gives

$$\alpha \begin{bmatrix} 1 \\ 0 \\ 0 \\ -1.5 \\ 0.5 \\ 0.5 \end{bmatrix} + \beta \begin{bmatrix} 0 \\ 1 \\ 0 \\ -0.5 \\ -0.5 \\ -0.5 \end{bmatrix} + \gamma \begin{bmatrix} 0 \\ 0 \\ 1 \\ 0.5 \\ -0.5 \\ -0.5 \end{bmatrix} = \begin{bmatrix} 0 \\ 0 \\ 0 \\ 0 \\ 0 \\ 0 \end{bmatrix}$$

For Π_1, Π_2, and Π_3 to be independent, α, β, and γ must both be zero. Solving the above matrix equation yields

$$\alpha + 0 + 0 = 0$$
$$0 + \beta + 0 = 0$$
$$0 + 0 + \gamma = 0$$
$$-1.5\alpha - 0.5\beta + 0.5\gamma = 0$$
$$0.5\alpha + 0.5\beta - 0.5\gamma = 0$$
$$0.5\alpha + 0.5\beta - 0.5\gamma = 0$$

The above system of linear equations shows that $\alpha = \beta = \gamma = 0$; thus, Π_1, Π_2, and Π_3 are independent of each other.

The functional solution is

$$f(\Pi_1, \Pi_2, \Pi_3) = f\left(t\sqrt{\frac{GM_{Sun}}{L_{SunEarth}^3}}, \frac{A}{\sqrt{GL_{SunEarth}M_{Sun}}}, v\sqrt{\frac{L_{SunEarth}}{GM_{Sun}}} \right) = 0$$

or, in terms of A

$$\frac{A}{\sqrt{GL_{SunEarth}M_{Sun}}} = \kappa \times f(\Pi_1, \Pi_3)$$

Let us assume $f(\Pi_1, \Pi_3)$ is a power law function, then

$$\frac{A}{\sqrt{GL_{SunEarth}M_{Sun}}} = \kappa \times \Pi_1^\alpha \Pi_2^\beta = \kappa \times t\sqrt{\frac{GM_{Sun}}{L_{SunEarth}^3}}^\alpha \left(v\sqrt{\frac{L_{SunEarth}}{GM_{Sun}}} \right)^\beta$$

Now, we use our physical knowledge to simplify the above equation. We said the initial condition for our demonstration is $t = 0$. If $\alpha > 0$ at $t = 0$, then $A = 0$, which means Earth is stationary—an unlikely consequence. If $\alpha < 0$ at $t = 0$, then A approaches infinity—another unlikely consequence. Thus, $\alpha = 0$ and the above equation becomes

$$\frac{A}{\sqrt{GL_{\text{SunEarth}}M_{\text{Sun}}}} = \kappa \times \left(v\sqrt{\frac{L_{\text{SunEarth}}}{GM_{\text{Sun}}}} \right)^{\beta}$$

$$A = \kappa\sqrt{GL_{\text{SunEarth}}M_{\text{Sun}}} \left(v\sqrt{\frac{L_{\text{SunEarth}}}{GM_{\text{Sun}}}} \right)^{\beta}$$

Note that A is independent of t and that all the terms to the right of the equality sign are constants, thereby "proving" Kepler's Second Law.

5.8 FLUID FLOW IN A CIRCULAR PIPE

All chemical engineers at some point in their careers become interested in fluid flow in circular pipes. The variable of most concern is pressure drop between flow initiation and flow termination. In petrochemical plants and along transport pipelines, there is only so much pressure available to induce flow. In other words, pressure loss must be minimized.

The geometric parameters for pipe flow are: pipe diameter D [L], pipe length l [L], and pipe wall roughness r [L]. The material parameters are fluid density ρ [L^{-3}M] and fluid viscosity μ [L^{-1}MT^{-1}]. The process parameters are fluid velocity v [LT^{-1}] and pressure drop per pipe length dP/dl [L^{-2}MT^{-2}].

The Dimension Table is

Variables		dP/dl	D	l	r	v	ρ	μ
Dimensions	L	-2	1	1	1	1	-3	-1
	M	1	0	0	0	0	1	1
	T	-2	0	0	0	-1	0	-1

and the Dimension matrix is

$$\begin{bmatrix} -2 & 1 & 1 & 1 & 1 & -3 & -1 \\ 1 & 0 & 0 & 0 & 0 & 1 & 1 \\ -2 & 0 & 0 & 0 & -1 & 0 & -1 \end{bmatrix}$$

The Rank matrix is the largest, nonsingular, square matrix contained in the Dimension matrix. Since the Dimension matrix has three rows, the largest square matrix will be a 3×3 matrix. Starting with the rightmost column of the Dimension matrix, the Rank matrix is

$$R = \begin{bmatrix} 1 & -3 & -1 \\ 0 & 1 & 1 \\ -1 & 0 & -1 \end{bmatrix}$$

and its determinant, calculated by a free-for-use matrix calculator on the Internet, is

$$|R| = \begin{vmatrix} 1 & -3 & -1 \\ 0 & 1 & 1 \\ -1 & 0 & -1 \end{vmatrix} = 3$$

Thus, the determinant is nonsingular and the number of dimensionless parameters is

$$N_P = N_{Var} - R = 7 - 3 = 4$$

The inverse of the Rank matrix is

$$R^{-1} = \begin{bmatrix} 1 & -3 & -1 \\ 0 & 1 & 1 \\ -1 & 0 & -1 \end{bmatrix}^{-1} = \begin{bmatrix} 1 & 3 & 2 \\ 1 & 2 & 1 \\ -1 & -3 & -1 \end{bmatrix}$$

and the Bulk matrix is

$$B = \begin{bmatrix} -2 & 1 & 1 & 1 \\ 1 & 0 & 0 & 0 \\ -2 & 0 & 0 & 0 \end{bmatrix}$$

Multiplying $-R^{-1}$ and B yields

$$-R^{-1} \times B = - \begin{bmatrix} 1 & 3 & 2 \\ 1 & 2 & 1 \\ -1 & -3 & -1 \end{bmatrix} \begin{bmatrix} -2 & 1 & 1 & 1 \\ 1 & 0 & 0 & 0 \\ -2 & 0 & 0 & 0 \end{bmatrix}$$

$$= \begin{bmatrix} -3 & 1 & 1 & 1 \\ -2 & 1 & 1 & 1 \\ 1 & -1 & -1 & -1 \end{bmatrix}$$

Since there are no zeroes in $-R^{-1} \times B$, there will be no zeroes in $(-R^{-1} \times B)^{\mathrm{T}}$. Thus, there are no irrelevant variables in this example.

The Total matrix is, then

$$
\begin{array}{c}
\qquad \Pi_1 \quad \Pi_2 \quad \Pi_3 \quad \Pi_4 \\
\begin{array}{c}
\mathrm{d}P/\mathrm{d}l \\
D \\
l \\
r \\
v \\
\rho \\
\mu
\end{array}
\left[
\begin{array}{ccccccc}
1 & 0 & 0 & 0 & 0 & 0 & 0 \\
0 & 1 & 0 & 0 & 0 & 0 & 0 \\
0 & 0 & 1 & 0 & 0 & 0 & 0 \\
0 & 0 & 0 & 1 & 0 & 0 & 0 \\
-3 & 1 & 1 & 1 & -1 & -3 & -2 \\
-2 & 1 & 1 & 1 & -1 & -2 & -1 \\
1 & -1 & -1 & -1 & 1 & 3 & 1
\end{array}
\right]
\end{array}
$$

This example has four dimensionless parameters; therefore, the Unit matrix has four columns, which are identified by Π_1, Π_2, Π_3, and Π_4. Matching variables with column elements in the Total matrix gives us the dimensionless parameters, which are

$$
\Pi_1 = \frac{(\mathrm{d}P/\mathrm{d}l)\mu}{v^3 \rho^2}; \quad \Pi_2 = \frac{Dv\rho}{\mu} = \frac{\rho Dv}{\mu}; \quad \Pi_3 = \frac{lv\rho}{\mu}; \quad \text{and } \Pi_4 = \frac{rv\rho}{\mu}
$$

Note that Π_2 is the Reynolds number.

So, are these four dimensionless parameters independent of each other? The test for linear independence is

$$
\alpha\Pi_1 + \beta\Pi_2 + \gamma\Pi_3 + \delta\Pi_4 = 0
$$

Substituting the matrix Π_1 and Π_2 columns into the above equation gives

$$
\alpha
\begin{bmatrix}
1 \\ 0 \\ 0 \\ 0 \\ -3 \\ -2 \\ 1
\end{bmatrix}
+ \beta
\begin{bmatrix}
0 \\ 1 \\ 0 \\ 0 \\ 1 \\ 1 \\ -1
\end{bmatrix}
+ \gamma
\begin{bmatrix}
0 \\ 0 \\ 1 \\ 0 \\ 1 \\ 1 \\ -1
\end{bmatrix}
+ \delta
\begin{bmatrix}
0 \\ 0 \\ 0 \\ 1 \\ 1 \\ 1 \\ -1
\end{bmatrix}
=
\begin{bmatrix}
0 \\ 0 \\ 0 \\ 0 \\ 0 \\ 0 \\ 0
\end{bmatrix}
$$

For Π_1, Π_2, Π_3, and Π_4 to be independent, α, β, γ, and δ must all be zero. Solving the above matrix equation yields

$$\alpha + 0 + 0 + 0 = 0$$
$$0 + \beta + 0 + 0 = 0$$
$$0 + 0 + \gamma + 0 = 0$$
$$0 + 0 + 0 + \delta = 0$$
$$-3\alpha + \beta + \gamma + \delta = 0$$
$$-2\alpha + \beta + \gamma + \delta = 0$$
$$\alpha - \beta - \gamma - \delta = 0$$

The above system of linear equations shows that $\alpha = \beta = \gamma = \delta = 0$; thus, Π_1, Π_2, Π_3, and Π_4 are independent of each other.

Since these dimensionless parameters are independent of each other, we can combine them to clarify the physics of the result. Doing so yields three different dimensionless parameters, which are

$$\frac{\Pi_3}{\Pi_2} = \Pi_{3/2} = \frac{lv\rho/\mu}{Dv\rho/\mu} = \frac{1}{D}$$

$$\frac{\Pi_4}{\Pi_2} = \Pi_{4/2} = \frac{rv\rho/\mu}{Dv\rho/\mu} = \frac{r}{D}$$

$$\Pi_1\Pi_2 = \Pi_{12} = \left(\frac{(dP/dl)\mu}{v^3\rho^2}\right)\left(\frac{Dv\rho}{\mu}\right) = \frac{D(dP/dl)}{\rho v^2} = 2f$$

where $2f$ is the Fanning friction number. The solution for this example, in functional notation, is

$$f(\Pi_{12}, \Pi_{3/2}, \Pi_{4/2}, \Pi_2) = f\left(\frac{D(dP/dl)}{\rho v^2}, \frac{1}{D}, \frac{r}{D}, \frac{\rho Dv}{\mu}\right) = 0$$

Writing the function in terms of Π_{12} gives us

$$\Pi_{12} = 2f = \kappa \times f(\Pi_{3/2}, \Pi_{4/2}, \Pi_2) = \kappa \times f\left(\frac{1}{D}, \frac{r}{D}, \frac{\rho Dv}{\mu}\right)$$

or

$$\frac{D(dP/dl)}{\rho v^2} = 2f = \kappa \times f(\Pi_{3/2}, \Pi_{4/2}, \Pi_2)$$

where κ and $f(\Pi_{3/2}, \Pi_{4/2}, \Pi_2)$ are determined by experimentation. The most common plots are $2f$ versus Π_2, i.e., the Reynolds number, with

r/D, i.e., $\Pi_{4/2}$, as the parametric line. A second plot would be $2f$ versus Π_2 with $\Pi_{3/2}$, i.e., l/D, as the parametric line. Such plots are common in the engineering literature.[3-5]

In this example, we have three independent dimensionless parameters. The total number of experiments required to define this function is

$$N_{\text{Expts}}^{\text{Total}} = N_{\text{Expts}}^{M_{\text{IndepVar}}} = 5^3 = 125$$

If we solve this example dimensionally, the solution function is

$$\frac{dP}{dZ} = f(D, l, r, v, \rho, \mu)$$

$N_{\text{Expts}}^{\text{Total}}$ for the dimensional solution function is

$$N_{\text{Expts}}^{\text{Total}} = N_{\text{Expts}}^{M_{\text{IndepVar}}} = 5^6 = 15,625$$

The difference between 125 experiments and 15,625 is the career lifetimes of several researchers. If we assume 250 working days per year and one experiment performed per day, then 15,625 experiments would require one researcher 62.5 years of effort; two researchers require 31.3 years; three researchers require 20.8 years; four researchers require 15.6 years, and so on. Conversely, if the project is to be completed in 3 years, then it requires 21 researchers. If each researcher is valued at $150,000 per year, a conservative estimate, then the project requires $3.2 million for personnel only. The dimensionless solution requires 125 experiments or half a year to complete or $75,000 for personnel.

5.9 SUMMARY

In this chapter, we demonstrated the use of Dimensional Analysis via its matrix formulation. Our first two examples each generated one dimensionless parameter. They constitute the simplest form of Dimensional Analysis. The third, fourth, and fifth examples each generated two dimensionless parameters. Thus, we had to determine that the two dimensionless parameters were linearly independent of each other. We established their linear independence using linear algebra techniques. Our sixth example generated three dimensionless parameters, which we determined were linearly independent of each other.

Our seventh example generated four dimensionless parameters; again, they were shown to be linearly independent of each other.

From these examples, we conclude: the matrix formulation of Dimensional Analysis ensures that the resulting dimensionless parameters are linearly independent of each other.

With each set of examples, we implemented a simplification that increases the rate at which we arrive at a solution to our problem.

We also calculated how many experiments are required to determine the function for the Dimensional Analysis result and for the dimensional result. In every case, the former is much smaller than the latter; thus, project development time is shorter and costs less when using Dimensional Analysis for project development than using dimensional methods.

REFERENCES

1. H. Huntley, *Dimensional Analysis*, Dover Publications, Inc., New York, NY, 1967.

2. T. Szirtes, *Dimensional Analysis and Modeling*, Butterworth-Heinemann, Burlington, MA, 2007.

3. L. Moody, *Transactions of the American Society of Mechanical Engineers*, 66, 671 (1944).

4. J. Hunsaker, B. Rightmire, *Engineering Applications of Fluid Mechanics*, McGraw-Hill Book Company, Inc., New York, NY, 1947, pp. 126–127.

5. G. Murphy, *Similitude in Engineering*, The Ronald Press Company, New York, NY, 1950, p. 140.

CHAPTER 6

Thermal Examples of Dimensional Analysis

6.1 INTRODUCTION

Energy, or heat, transfer requires a fourth fundamental dimension. That fourth fundamental dimension is heat, which is the transfer of energy from a region of higher temperature to a region of lower temperature; thus, the fourth dimension could be temperature or it could be heat. The SI system of units uses temperature as a fundamental dimension. In most cases, we will do the same.

6.2 IDEAL GAS LAW[1(PP118−119)]

Imagine yourself living in county Kent, southeast of London, during the first decade of the nineteenth century. Your vocation is country squire, but your avocation is natural philosophy. Every couple of months you ride to London to attend a meeting of the Royal Society where the latest natural philosophy is presented. Upon receipt of the Society's *Proceedings*, you read it avidly. Of late, you have become interested in the physical behavior of gases. You know Boyle's Law, which states that $PV = \kappa_B$, where P is pressure, V is volume, and κ_B is a constant. You also know Charles' Law and Guy Lussac's Law, which state that, when all other process variables are held constant, $V = \kappa_{CGL} \times T$, where, again, V is volume, T is temperature, and κ_{CGL} is another constant. You have just learned in the latest *Proceedings* that Amedeo Avogadro has hypothesized that for dilute gases, equal volumes contain equal numbers of gas molecules. One day, after a hearty breakfast, you retire to your library with the intention of summarizing these gas laws in one mathematical expression.

You begin by considering a cylinder of gas at a given temperature. You assume the gas molecule has mass and that its collision with the cylinder wall is perfectly elastic. Pressure, you assume, is the reversal of momentum that occurs when the gas molecules rebound from colliding with the cylinder wall. You also assume that the size of the gas molecule is negligible compared to the radius of the cylinder. Your question is: what is the relationship between pressure and the other variables of the process?

There are no geometric variables. The material variable is molecular mass m [M]. The process variables are absolute temperature K [θ], pressure P [$L^{-1}MT^{-2}$], and the number of gas molecules per unit volume N [L^{-3}]. You realize you need a dimensional constant for this problem; it is the gas constant per unit mass R [$L^2T^{-2}\theta^{-1}$].

Your Dimension Table is

Variables		m	K	N	P	R
Dimensions	L	0	0	-3	-1	2
	M	1	0	0	1	0
	T	0	0	0	-2	-2
	θ	0	0	1	0	-1

Thus, you write the Dimension matrix as

$$\begin{bmatrix} 0 & 0 & -3 & -1 & 2 \\ 1 & 0 & 0 & 1 & 0 \\ 0 & 0 & 0 & -2 & -2 \\ 0 & 1 & 0 & 0 & -1 \end{bmatrix}$$

The largest square matrix within the Dimension matrix is 4×4. Starting from the rightmost column, the Rank matrix is

$$R = \begin{bmatrix} 0 & -3 & -1 & 2 \\ 0 & 0 & 1 & 0 \\ 0 & 0 & -2 & -2 \\ 1 & 0 & 0 & -1 \end{bmatrix}$$

and its determinant is

$$|R| = \begin{vmatrix} 0 & -3 & -1 & 2 \\ 0 & 0 & 1 & 0 \\ 0 & 0 & -2 & -2 \\ 1 & 0 & 0 & -1 \end{vmatrix} = -6$$

Thus, the Rank of the Dimension matrix is 4. The number of dimensionless parameters is

$$N_P = N_{Var} - R = 5 - 4 = 1$$

The inverse of the Rank matrix is

$$R^{-1} = \begin{bmatrix} 0 & -3 & -1 & 2 \\ 0 & 0 & 1 & 0 \\ 0 & 0 & -2 & -2 \\ 1 & 0 & 0 & -1 \end{bmatrix} = \begin{bmatrix} 0 & -1 & -0.5 & 1 \\ -0.3 & -1 & -0.3 & 0 \\ 0 & 1 & 0 & 0 \\ 0 & -1 & -0.5 & 0 \end{bmatrix}$$

You identify the Bulk matrix as

$$B = \begin{bmatrix} 0 \\ 1 \\ 0 \\ 0 \end{bmatrix}$$

and $-R^{-1} \times B$ is

$$-R^{-1} \times B = - \begin{bmatrix} 0 & -1 & -0.5 & 1 \\ -0.3 & -1 & -0.3 & 0 \\ 0 & 1 & 0 & 0 \\ 0 & -1 & -0.5 & 0 \end{bmatrix} \begin{bmatrix} 0 \\ 1 \\ 0 \\ 0 \end{bmatrix} = \begin{bmatrix} 1 \\ 1 \\ -1 \\ 1 \end{bmatrix}$$

Since there are no rows entirely of zero in $-R^{-1} \times B$, there will be no columns entirely of zero in $(-R^{-1} \times B)^T$. Thus, there are no irrelevant variables in this example.

You next write your Total matrix as

$$T = \begin{matrix} m \\ K \\ N \\ P \\ R \end{matrix} \begin{matrix} \Pi_1 \\ \begin{bmatrix} 1 & 0 & 0 & 0 & 0 \\ 1 & 0 & -1 & -0.5 & 1 \\ 1 & -0.3 & -1 & -0.3 & 0 \\ -1 & 0 & 1 & 0 & 0 \\ 1 & 0 & -1 & -0.5 & 0 \end{bmatrix} \end{matrix}$$

To obtain the dimensionless parameter, you assign the power for each variable, listed along the left side of the Total matrix, from the appropriate element of column Π_1. Your dimensionless parameter is

$$\Pi_1 = \frac{mNRK}{P}$$

You, therefore, write your solution in functional notation as

$$f(\Pi_1) = f\left(\frac{mNRK}{P}\right) = 0$$

But, you realize the function can be written as

$$\frac{mNRK}{P} = \kappa$$

Rearranging the above equation yields

$$RK = \left(\frac{\kappa}{mN}\right) \times P$$

But, mN is the weight of an individual gas molecule times the number of gas molecules per unit volume; thus, the above equation becomes

$$RK = \kappa VP$$

where you identify κ as the inverse of the number of gas "moles." In other words,

$$nRK = PV$$

You immediately write to the Royal Society about your result.

6.3 FLOWING HOT WATER HEATER (BOUSSINESQ'S PROBLEM)[1(PP124—126),2(PP551—555),3(PP95—97),4]

You live in a hard water area of the United States where deposit accumulates in your hot water heater. The deposit develops from heating the hard water, which causes gypsum to precipitate from it. You decide that the solution is a hot water heater that heats and flushes the hot zone simultaneously, thereby ensuring a deposit-free heating zone. So, you sketch a pipe through which water flows on demand over a heating element. The heating element transmits energy, in the form of heat, to the water and the flowing water flushes any micro-precipitant downstream. The question is: how does heat transmission relate to the other variables of your invention?

The geometric variable is the size of the thermal conductor, i.e., the heating element L [L]. The material variables are water heat capacity C_P [$L^{-1}MT^{-2}\theta^{-1}$] and water heat conductivity k [$L^2MT^{-3}\theta^{-1}$]. The process parameters are heat transferred q [L^2MT^{-3}], water flow velocity v [LT^{-1}], and temperature difference between the heating element and the water ΔK [$\Delta\theta$].

Your Dimension Table is

Variables		L	ΔK	v	q	C_P	k
Dimensions	L	1	0	1	2	−1	1
	M	0	0	0	1	1	1
	T	0	0	−1	−3	−2	−3
	θ	0	0	1	0	0	−1

The Dimension matrix is

$$\begin{bmatrix} 1 & 0 & 1 & 2 & -1 & 1 \\ 0 & 0 & 0 & 1 & 1 & 1 \\ 0 & 0 & -1 & -3 & -2 & -3 \\ 0 & 1 & 0 & 0 & -1 & -1 \end{bmatrix}$$

You identify the largest square matrix in the Dimension matrix as

$$R = \begin{bmatrix} 1 & 2 & -1 & 1 \\ 0 & 1 & 1 & 1 \\ -1 & -3 & -2 & -3 \\ 0 & 0 & -1 & -1 \end{bmatrix}$$

and then, you calculate the determinant of R, which is

$$|R| = \begin{vmatrix} 1 & 2 & -1 & 1 \\ 0 & 1 & 1 & 1 \\ -1 & -3 & -2 & -3 \\ 0 & 0 & -1 & -1 \end{vmatrix} = 1$$

Thus, the Rank of the Dimension matrix is 4. The number of dimensionless variables is

$$N_P = N_{Var} - R = 6 - 4 = 2$$

Using a free-for-use matrix calculator on the Internet, you find the inverse of the Rank matrix to be

$$R^{-1} = \begin{bmatrix} 1 & 2 & -1 & 1 \\ 0 & 1 & 1 & 1 \\ -1 & -3 & -2 & -3 \\ 0 & 0 & -1 & -1 \end{bmatrix}^{-1} = \begin{bmatrix} -1 & -4 & -2 & 1 \\ 0 & 1 & 0 & 1 \\ -1 & -1 & -1 & 1 \\ 1 & 1 & 1 & -2 \end{bmatrix}$$

Your Bulk matrix is

$$B = \begin{bmatrix} 1 & 0 \\ 0 & 0 \\ 0 & 0 \\ 0 & 1 \end{bmatrix}$$

and $-R^{-1} \times B$ is

$$-R^{-1} \times B = - \begin{bmatrix} -1 & -4 & -2 & 1 \\ 0 & 1 & 0 & 1 \\ -1 & -1 & -1 & 1 \\ 1 & 1 & 1 & -2 \end{bmatrix} \begin{bmatrix} 1 & 0 \\ 0 & 0 \\ 0 & 0 \\ 0 & 1 \end{bmatrix} = \begin{bmatrix} 1 & -1 \\ 0 & -1 \\ 1 & -1 \\ -1 & 2 \end{bmatrix}$$

Calculating matrix V to determine if your analysis includes any irrelevant variables, you obtain, with variable identification,

$$V = (-R^{-1} \times B)^{\mathrm{T}} = \begin{bmatrix} 1 & -1 \\ 0 & -1 \\ 1 & -1 \\ -1 & 2 \end{bmatrix}^{\mathrm{T}} = \begin{array}{cccc} v & q & C_{\mathrm{P}} & k \\ \begin{bmatrix} 1 & 0 & 1 & -1 \\ -1 & -1 & -1 & 2 \end{bmatrix} \end{array}$$

No column of matrix V is entirely comprised of zeroes; thus, you have no irrelevant variables in this example.

Your Total matrix is

$$T = \begin{array}{c} \\ L \\ \Delta K \\ v \\ q \\ C_{\mathrm{P}} \\ k \end{array} \begin{array}{cc} \Pi_1 & \Pi_2 \\ \begin{bmatrix} 1 & 0 & 0 & 0 & 0 & 0 \\ 0 & 1 & 0 & 0 & 0 & 0 \\ 1 & -1 & -1 & -4 & -2 & 1 \\ 0 & -1 & 0 & 1 & 0 & 1 \\ 1 & -1 & -1 & -1 & -1 & 1 \\ -1 & 2 & 1 & 1 & 1 & -2 \end{bmatrix} \end{array}$$

The dimensionless parameters, reading down the Π_i columns of the Total matrix, are

$$\Pi_1 = \frac{L v C_{\mathrm{P}}}{k} \quad \text{and} \quad \Pi_2 = \frac{k^2 \Delta \Theta}{v q C_{\mathrm{P}}}$$

You want q to be your dependent variable, but q is in the denominator of Π_2. That is not an optimal location for a dependent variable. However, inverting Π_2 moves q into the numerator, a more appropriate location for a dependent variable. You also note that inverting Π_1, then multiplying it with Π_2^{-1} produces a simpler dimensionless parameter, namely

$$\Pi_1^{-1}\Pi_2^{-1} = \left(\frac{k}{LvC_P}\right)\left(\frac{qvC_P}{k^2\Delta K}\right) = \frac{q}{kL\Delta K} = \Pi_3$$

Therefore, the solution in functional notation is

$$f(\Pi_1, \Pi_3) = 0$$

or

$$\Pi_3 = \kappa \times f(\Pi_1)$$

The total number of experiments $N_{\text{Expts}}^{\text{Total}}$ required to define the above function is

$$N_{\text{Expts}}^{\text{Total}} = 5^1 = 5$$

You wonder how many experiments you would have to perform if you had analyzed your invention dimensionally. Thus, you write the dimensional solution for your invention as

$$q = \kappa \times f(L, \Delta K, v, C_P, k)$$

The required experiments to define this new function is

$$N_{\text{Expts}}^{\text{Total}} = 5^5 = 3125$$

Comparing the number of experiments required for your project, you opt to validate your invention using the dimensionless solution. After experimentally validating your invention, you discuss it with your attorney and learn that someone else has already patented your idea. Thus, another lesson learned at the University of Life: check with your patent attorney before commencing experimental work to confirm an idea.

6.4 FILM COEFFICIENT FOR BATCH MIXER HEAT TRANSFER[1(PP122−124),2(PP535−537)]

At some point in our careers as chemical engineers, we will operate a batch mixer. From a safety viewpoint, we are concerned about removing the heat that may occur upon mixing two liquids. We want to avoid thermal "runaway" when adding one fluid to another fluid control, upon fluid mixing. From a production capacity viewpoint, we are concerned with applying heat fast enough to initiate reaction in a reasonable amount of time. Therefore, consider a reactor with a coiled pipe placed internally to add or remove heat from the mixing liquids. The reactor is equipped with a turbine agitator that moves the contained liquids quickly enough that heat transfer within the mixing liquids is not the limiting factor. The rate at which fluid flows through the internal coil establishes a temperature difference between the inside and outside of the internal coil. As with any fluid flowing over a surface, a thin film forms along the internal surface of the coil. Heat must "diffuse" through this film when being transferred into or from the reactor. Thus, the rate of heat transfer inside the coil depends upon the thickness of this film, and film thickness depends upon fluid viscosity, fluid velocity, and coil diameter. We define the heat transfer film coefficient as

$$h = \frac{q}{At(\Delta K)}$$

where q is the heat transferred into or from the the reactor [L^2MT^{-2}], A is the area through which heat transfers [L^2], t is time [T], and ΔK is temperature difference [θ]. Thus, the heat transfer film coefficient has dimensions of [$MT^{-3}\theta^{-1}$].

The question is: what is the relationship of h to the various operating variables of the reactor? The geometric variable is pipe diameter D [L]. The material variables are fluid heat capacity C_P [$L^2T^{-2}\theta^{-1}$], fluid heat conductivity k [$LMT^{-3}\theta^{-1}$], and dynamic viscosity μ [$L^{-1}MT^{-1}$]. The process variables are heat transfer film coefficient h [$MT^{-3}\theta^{-1}$] and mass rate of fluid per unit area Q [$L^{-2}MT^{-1}$]. The Dimension Table is

Variables		h	Q	C_P	D	k	μ
Dimensions	L	0	−2	2	1	1	−1
	M	1	1	0	0	1	1
	T	−3	−1	−2	0	−3	−1
	θ	−1	0	−1	0	−1	0

The Dimension matrix is

$$\begin{bmatrix} 0 & -2 & 2 & 1 & 1 & -1 \\ 1 & 1 & 0 & 0 & 1 & 1 \\ -3 & -1 & -2 & 0 & -3 & -1 \\ -1 & 0 & -1 & 0 & -1 & 0 \end{bmatrix}$$

The largest square matrix for this Dimension matrix is 4×4, which is as follows:

$$R = \begin{bmatrix} 2 & 1 & 1 & -1 \\ 0 & 0 & 1 & 1 \\ -2 & 0 & -3 & -1 \\ -1 & 0 & -1 & 0 \end{bmatrix}$$

The determinant of R is

$$|R| = \begin{vmatrix} 2 & 1 & 1 & -1 \\ 0 & 0 & 1 & 1 \\ -2 & 0 & -3 & -1 \\ -1 & 0 & -1 & 0 \end{vmatrix} = 0$$

Thus, R cannot be a 4×4 matrix. So, what does this mean? It means we have at least one irrelevant fundamental dimension in our analysis. We must delete one fundamental dimension from our analysis. We will delete temperature θ. The new Dimensional Table is

Variables		h	Q	C_P	D	k	μ
Dimensions	L	0	-2	2	1	1	-1
	M	1	1	0	0	1	1
	T	-3	-1	-2	0	-3	-1

and the new Dimension matrix is

$$\begin{bmatrix} 0 & -2 & 2 & 1 & 1 & -1 \\ 1 & 1 & 0 & 0 & 1 & 1 \\ -3 & -1 & -2 & 0 & -3 & -1 \end{bmatrix}$$

The largest square matrix for the new Dimension matrix is 3×3, which is as follows:

$$R = \begin{bmatrix} 1 & 1 & -1 \\ 0 & 1 & 1 \\ 0 & -3 & -1 \end{bmatrix}$$

Its determinant is

$$|R| = \begin{vmatrix} 1 & 1 & -1 \\ 0 & 1 & 1 \\ 0 & -3 & -1 \end{vmatrix} = 2$$

Thus, the Rank of the new Dimension matrix is 3. The number of dimensionless parameters is

$$N_P = N_{Var} - R = 6 - 3 = 3$$

The inverse of the Rank matrix is

$$R^{-1} = \begin{bmatrix} 1 & 1 & -1 \\ 0 & 1 & 1 \\ 0 & -3 & -1 \end{bmatrix}^{-1} = \begin{bmatrix} 1 & 2 & 1 \\ 0 & -0.5 & -0.5 \\ 0 & 1.5 & 0.5 \end{bmatrix}$$

and the Bulk matrix is

$$B = \begin{bmatrix} 0 & -2 & 2 \\ 1 & 1 & 0 \\ -3 & -1 & -2 \end{bmatrix}$$

Therefore, $-R^{-1} \times B$ is

$$-R^{-1} \times B = -\begin{bmatrix} 1 & 2 & 1 \\ 0 & -0.5 & -0.5 \\ 0 & 1.5 & 0.5 \end{bmatrix}\begin{bmatrix} 0 & -2 & 2 \\ 1 & 1 & 0 \\ -3 & -1 & -2 \end{bmatrix} = \begin{bmatrix} 1 & 1 & 0 \\ -1 & 0 & -1 \\ 0 & -1 & 1 \end{bmatrix}$$

Calculating matrix V to determine if our analysis includes any irrelevant variables, we obtain, with variable identification,

$$(-R^{-1} \times B)^T = \begin{bmatrix} 1 & 1 & 0 \\ -1 & 0 & -1 \\ 0 & -1 & 1 \end{bmatrix}^T = \overset{\begin{matrix} D & k & \mu \end{matrix}}{\begin{bmatrix} 1 & -1 & 0 \\ 1 & 0 & -1 \\ 0 & -1 & 1 \end{bmatrix}}$$

No column of matrix V is entirely comprised of zeroes; thus, we have no irrelevant variables in this example.

The Total matrix is

$$T = \begin{matrix} h \\ Q \\ C_P \\ D \\ k \\ \mu \end{matrix} \overset{\begin{matrix} \Pi_1 & \Pi_2 & \Pi_3 \end{matrix}}{\begin{bmatrix} 1 & 0 & 0 & 0 & 0 & 0 \\ 0 & 1 & 0 & 0 & 0 & 0 \\ 0 & 0 & 1 & 0 & 0 & 0 \\ 1 & 1 & 0 & 0 & -2 & 2 \\ -1 & 0 & -1 & 1 & 1 & 0 \\ 0 & -1 & 1 & -3 & -1 & -2 \end{bmatrix}}$$

The dimensionless parameters, reading down each Π_i column of the Total matrix, are

$$\Pi_1 = \frac{hD}{k}, \quad \Pi_2 = \frac{QD}{\mu}, \quad \Pi_3 = \frac{\mu C_P}{k}$$

Note that Π_1 is the Nusselt number, Π_3 is the Prandtl number, and Π_2 is one of the variant forms of the Reynolds number.

The solution in functional form is

$$f(\Pi_1, \Pi_2, \Pi_3) = f\left(\frac{hD}{k}, \frac{QD}{\mu}, \frac{\mu C_P}{k}\right) = 0$$

and in terms of Π_1, the functional solution is

$$\Pi_1 = \kappa \times f(\Pi_2, \Pi_3) = \kappa \times f\left(\frac{QD}{\mu}, \frac{\mu C_P}{k}\right) = 0$$

Thus, the dimensionless solution requires 25 experiments while the dimensional solution

$$h = f(Q, C_P, D, k, \mu)$$

requires 3125 experiments to define h. The dimensionless solution represents a significant cost savings with respect to research and engineering effort. Also, this example shows that the Dimensional Analysis result can be displayed on one plot. The experimental data for the dimensional solution requires a book.

6.5 STEADY STATE HEAT TRANSFER IN BUBBLE COLUMNS[3(PP185−189)]

Consider a cylindrical column into which liquid and gas are pumped through bottom nozzles. Both fluids flow upward to the discharge port at the top of the column. At steady state, how does the heat transfer coefficient relate to the other pertinent variables of the process?

The geometric variable is column diameter D [L]. The material variables are liquid density ρ [$L^{-3}M$], liquid heat capacity C_P [$L^2MT^{-2}\theta^{-1}$], and liquid thermal conductivity k [$LMT^{-3}\theta^{-1}$]. The process variables are gas volumetric flow rate Q [L^3T^{-1}] and gravity difference between the two fluids $g\Delta\rho$ [$L^{-1}MT^{-2}$].

The Dimension Table is

Variables		h	D	ρ	μ	C_P	k	Q	$g\Delta\rho$
Dimensions	L	0	1	-3	-1	2	1	3	-2
	M	1	0	1	1	1	1	0	1
	T	-3	0	0	-1	-2	-3	-1	-2
	θ	-1	0	0	0	-1	-1	0	0

and the Dimension matrix is

$$\begin{bmatrix} 0 & 1 & -3 & -1 & 2 & 1 & 3 & -2 \\ 1 & 0 & 1 & 1 & 1 & 1 & 0 & 1 \\ -3 & 0 & 0 & -1 & -2 & -3 & -1 & -2 \\ -1 & 0 & 0 & 0 & -1 & -1 & 0 & 0 \end{bmatrix}$$

The largest square matrix for this Dimension matrix is 4×4, which is

$$R = \begin{bmatrix} 2 & 1 & 3 & -2 \\ 1 & 1 & 0 & 1 \\ -2 & -3 & -1 & -2 \\ -1 & -1 & 0 & 0 \end{bmatrix}$$

Its determinant is

$$|R| = \begin{vmatrix} 2 & 1 & 3 & -2 \\ 1 & 1 & 0 & 1 \\ -2 & -3 & -1 & -2 \\ -1 & -1 & 0 & 0 \end{vmatrix} = -4$$

The Rank of this Dimension matrix is 4. The number of dimensionless parameters is

$$N_P = N_{Var} - R = 8 - 4 = 4$$

The inverse of R is

$$R^{-1} = \begin{bmatrix} 2 & 1 & 3 & -2 \\ 1 & 1 & 0 & 1 \\ -2 & -3 & -1 & -2 \\ -1 & -1 & 0 & 0 \end{bmatrix}^{-1} = \begin{bmatrix} 0.25 & 2 & 0.75 & 0 \\ -0.25 & -2 & -0.75 & -1 \\ 0.25 & 0 & -0.25 & 1 \\ 0 & 1 & 0 & 1 \end{bmatrix}$$

and the Bulk matrix is

$$
B = \begin{bmatrix} 0 & 1 & -3 & -1 \\ 1 & 0 & 1 & 1 \\ -3 & 0 & 0 & -1 \\ -1 & 0 & 0 & 0 \end{bmatrix}
$$

Therefore, $-R^{-1} \times B$ is

$$
-R^{-1} \times B = - \begin{bmatrix} 0.25 & 2 & 0.75 & 0 \\ -0.25 & -2 & -0.75 & -1 \\ 0.25 & 0 & -0.25 & 1 \\ 0 & 1 & 0 & 1 \end{bmatrix} \begin{bmatrix} 0 & 1 & -3 & -1 \\ 1 & 0 & 1 & 1 \\ -3 & 0 & 0 & -1 \\ -1 & 0 & 0 & 0 \end{bmatrix}
$$

$$
= \begin{bmatrix} 0.25 & -0.25 & -1.25 & -1 \\ -1.25 & 0.25 & 1.25 & 1 \\ 0.25 & -0.25 & 0.75 & 0 \\ 0 & 0 & -1 & -1 \end{bmatrix}
$$

Calculating matrix V to determine if our analysis includes any irrelevant variables, we obtain, with variable identification,

$$
(-R^{-1} \times B)^{\mathrm{T}} = \begin{bmatrix} 0.25 & -0.25 & -1.25 & -1 \\ -1.25 & 0.25 & 1.25 & 1 \\ 0.25 & -0.25 & 0.75 & 0 \\ 0 & 0 & -1 & -1 \end{bmatrix}^{\mathrm{T}} = \begin{array}{cccc} C_{\mathrm{P}} & k & q & g\Delta\rho \\ \begin{bmatrix} 0.25 & -1.25 & 0.25 & 0 \\ -0.25 & 0.25 & -0.25 & 0 \\ -1.25 & 1.25 & 0.75 & -1 \\ -1 & 1 & 0 & -1 \end{bmatrix} \end{array}
$$

No column of matrix V is entirely comprised of zeroes; thus, we have no irrelevant variables in this example.

The Total matrix is

	Π_1	Π_2	Π_3	Π_4				
h	1	0	0	0	0	0	0	0
D	0	1	0	0	0	0	0	0
ρ	0	0	1	0	0	0	0	0
μ	0	0	0	1	0	0	0	0
C_{P}	0.25	-0.25	-1.25	-1	0.25	12	0.75	0
k	-1.25	0.25	1.25	1	-0.25	-2	-0.75	-1
Q	0.25	-0.25	0.75	0	0.25	0	-0.25	1
$g\Delta\rho$	0	0	-1	-1	0	1	0	1

with $T =$ labeling the left side of the total matrix.

The dimensionless parameters, reading down each Π_i column of the Total matrix, are

$$\Pi_1 = \frac{hC_P^{0.25}Q^{0.25}}{k^{1.25}}, \quad \Pi_2 = \frac{Dk^{0.25}}{C_P^{0.25}Q^{0.25}}, \quad \Pi_3 = \frac{\rho k^{1.25}Q^{0.75}}{C_P^{1.25}g\Delta\rho}, \quad \Pi_4 = \frac{\mu k}{C_P g\Delta\rho}$$

It is best to use dimensionless parameters with whole number indices on their variables. Doing so makes it easier to interpret their physical meaning. We can achieve whole number indices on variables by combining the members of a given set of dimensionless parameters. For example, from the above set of dimensionless parameters, we can combine Π_2, Π_3, and Π_4 to obtain a variant of the Reynolds number, namely

$$\frac{\Pi_3}{\Pi_2\Pi_4} = \frac{\rho k^{1.25}Q^{0.75}/C_P^{1.25}g\Delta\rho}{(Dk^{0.25}/C_P^{0.25}Q^{0.25})(\mu k/C_P g\Delta\rho)} = \frac{\rho Q}{\mu D}$$

Similarly, combining Π_1 and Π_2 produces the Nusselt number

$$\Pi_1\Pi_2 = \left(\frac{hC_P^{0.25}Q^{0.25}}{k^{1.25}}\right)\left(\frac{Dk^{0.25}}{C_P^{0.25}Q^{0.25}}\right) = \frac{hD}{k}$$

Combining Π_2 and Π_3 gives us a variant of the Froude number, thus

$$\frac{\Pi_3}{\Pi_2^5} = \frac{\rho k^{1.25}Q^{0.75}/C_P^{1.25}g\Delta\rho}{(Dk^{0.25}/C_P^{0.25}Q^{0.25})^5} = \frac{\rho Q^2}{D^5 g\Delta\rho}$$

We can obtain a variant of the Prandtl by combining the dimensionless parameters, i.e.,

$$\frac{\Pi_4}{\Pi_2^3\Pi_3} = \frac{(\mu k/C_P g\Delta\rho)}{(Dk^{0.25}/C_P^{0.25}Q^{0.25})^3(\rho k^{1.25}Q^{0.75}/C_P^{1.25}g\Delta\rho)} = \frac{\mu C_P}{\rho k D^3} = \text{Pr}$$

The solution to this example in functional form is

$$f\left(\Pi_1\Pi_2, \frac{\Pi_3}{\Pi_2\Pi_4}, \frac{\Pi_3}{\Pi_2^5}, \frac{\Pi_4}{\Pi_2^3\Pi_3}\right) = f(\text{Nu, Re, Fr, Pr}) = 0$$

In terms of the heat transfer coefficient, the solution is

$$\text{Nu} = \kappa \times f(\text{Re, Fr, Pr})$$

The total number of experiments required to define the above function is

$$N_{\text{Expts}}^{\text{Total}} = 5^3 = 125$$

If we solve this example dimensionally, we obtain the functional solution:

$$h = \kappa \times f(D, \rho, \mu, C_P, k, Q, g\Delta\rho)$$

The required experiments to define the function is

$$N_{\text{Expts}}^{\text{Total}} = 5^7 = 78,125$$

which dramatically shows the benefits of developing a project using Dimensional Analysis compared to our normal process development procedures.

6.6 HEAT TRANSFER INSIDE TUBES[5]

Consider a fluid flowing inside a pipe or tube. The heat transferred to the flowing fluid is

$$q = hA\Delta K\Delta t$$

where q is the heat transferred to the flowing fluid, h is the convective heat transfer coefficient for the process, A is the area through which heat is transferred, ΔK is the temperature difference between the pipe's wall temperature and the average temperature of the flowing fluid, given by $K_{\text{Ave}} = (K_{\text{CenterLine}} - K_{\text{Wall}})/2$, and Δt is the time duration of heat flow.

The geometric variable is pipe diameter D [L]. The material variables are fluid density ρ [L^{-3}M], fluid dynamic viscosity μ [L^{-1}MT^{-1}], fluid heat capacity C_P [L^{2}MT$^{-2}\theta^{-1}$], fluid thermal conductivity k [LMT$^{-3}\theta^{-1}$], and fluid heat transfer coefficient h [MT$^{-3}\theta^{-1}$]. The process variables are fluid velocity v [LT^{-1}], average fluid temperature K_{Ave} [θ], and pipe wall temperature K [θ]. The Dimensional Table is

Variables		K_{Ave}	K_{Pipe}	D	v	μ	ρ	C_P	k	h
Dimensions	L	0	0	1	1	−1	−3	2	1	0
	M	0	0	0	0	1	1	0	1	1
	T	0	0	0	−1	−1	0	−2	−3	−3
	θ	1	1	0	0	0	0	−1	−1	−1

from which we write the Dimension matrix, which is

$$\begin{bmatrix} 0 & 0 & 1 & 1 & -1 & -3 & 2 & 1 & 0 \\ 0 & 0 & 0 & 0 & 1 & 1 & 0 & 1 & 1 \\ 0 & 0 & 0 & -1 & -1 & 0 & -2 & -3 & -3 \\ 1 & 1 & 0 & 0 & 0 & 0 & -1 & -1 & -1 \end{bmatrix}$$

The largest square matrix for this Dimension matrix is 4×4, which is as follows:

$$R = \begin{bmatrix} -3 & 2 & 1 & 0 \\ 1 & 0 & 1 & 1 \\ 0 & -2 & -3 & -3 \\ 0 & -1 & -1 & -1 \end{bmatrix}$$

Its determinant is

$$|R| = \begin{vmatrix} -3 & 2 & 1 & 0 \\ 1 & 0 & 1 & 1 \\ 0 & -2 & -3 & -3 \\ 0 & -1 & -1 & -1 \end{vmatrix} = -1$$

The determinant for the Rank matrix is nonsingular; thus, the Rank of this Dimension matrix is 4. The number of dimensionless parameters is

$$N_P = N_{Var} - R = 9 - 4 = 5$$

The inverse of R is

$$R^{-1} = \begin{bmatrix} -3 & 2 & 1 & 0 \\ 1 & 0 & 1 & 1 \\ 0 & -2 & -3 & -3 \\ 0 & -1 & -1 & -1 \end{bmatrix}^{-1} = \begin{bmatrix} 0 & 1 & 1 & -2 \\ 0 & 0 & 1 & -3 \\ 1 & 3 & 1 & 0 \\ -1 & -3 & -2 & 2 \end{bmatrix}$$

and the Bulk matrix is

$$B = \begin{bmatrix} 0 & 0 & 1 & 1 & -1 \\ 0 & 0 & 0 & 0 & 1 \\ 0 & 0 & 0 & -1 & -1 \\ 1 & 1 & 0 & 0 & 0 \end{bmatrix}$$

and $-R^{-1} \times B$ is

$$
-R^{-1} \times B = - \begin{bmatrix} 0 & 1 & 1 & -2 \\ 0 & 0 & 1 & -3 \\ 1 & 3 & 1 & 0 \\ -1 & -3 & -2 & 2 \end{bmatrix} \begin{bmatrix} 0 & 0 & 1 & 1 & -1 \\ 0 & 0 & 0 & 0 & 1 \\ 0 & 0 & 0 & -1 & -1 \\ 1 & 1 & 0 & 0 & 0 \end{bmatrix}
$$

$$
= \begin{bmatrix} 2 & 2 & 0 & 1 & 0 \\ 3 & 3 & 0 & 1 & 1 \\ 0 & 0 & -1 & 0 & -1 \\ -2 & -2 & 1 & -1 & 0 \end{bmatrix}
$$

Calculating matrix V to determine if our analysis includes any irrelevant variables, we obtain, with variable identification,

$$
(-R^{-1} \times B)^T = \begin{bmatrix} 2 & 2 & 0 & 1 & 0 \\ 3 & 3 & 0 & 1 & 1 \\ 0 & 0 & -1 & 0 & -1 \\ -2 & -2 & 1 & -1 & 0 \end{bmatrix}^T = \begin{array}{c} \begin{matrix} \rho & C_P & k & h \end{matrix} \\ \begin{bmatrix} 2 & 3 & 0 & -2 \\ 2 & 3 & 0 & -2 \\ 0 & 0 & -1 & 1 \\ 1 & 1 & 0 & -1 \\ 0 & 1 & -1 & 0 \end{bmatrix} \end{array}
$$

No column of matrix V is entirely comprised of zeroes; thus, we have no irrelevant variables in this example.

The Total matrix is

		Π_1	Π_2	Π_3	Π_4	Π_5				
	K_{Ave}	1	0	0	0	0	0	0	0	0
	K_{Pipe}	0	1	0	0	0	0	0	0	0
	D	0	0	1	0	0	0	0	0	0
	v	0	0	0	1	0	0	0	0	0
$T =$	μ	0	0	0	0	1	0	0	0	0
	ρ	2	2	0	1	0	0	1	1	-2
	C_P	3	3	0	1	1	0	0	1	-3
	k	0	0	-1	0	-1	1	3	1	0
	h	-2	-2	1	-1	0	-1	-3	-2	2

The dimensionless parameters, reading down each Π_i column of the Total matrix, are

$$\Pi_1 = \frac{K_{Ave}\rho^2 C_P^3}{h^2}, \quad \Pi_2 = \frac{K_{Pipe}\rho^2 C_P^3}{h^2}, \quad \Pi_3 = \frac{hD}{k} = Nu, \quad \Pi_4 = \frac{\rho v C_P}{h},$$

$$\Pi_5 = \frac{\mu C_P}{k} = Pr$$

Manipulating the above dimensionless parameters in order to remove indices >1 gives us

$$\frac{\Pi_1}{\Pi_2} = \frac{K_{Ave}\rho^2 C_P^3/h^2}{K_{Pipe}\rho^2 C_P^3/h^2} = \frac{K_{Ave}}{K_{Pipe}}$$

$$\frac{\Pi_3 \Pi_4}{\Pi_5} = \frac{(hD/k)(\rho v C_P/h)}{\mu C_P/k} = \frac{\rho D v}{\mu} = Re$$

$$\frac{\Pi_4^2}{\Pi_1} = \frac{(\rho v C_P/h)^2}{K_{Ave}\rho^2 C_P^3/h^2} = \frac{v^2}{K_{Ave} C_P}$$

The solution to this example in functional form is

$$f\left(\Pi_3, \Pi_5, \frac{\Pi_3 \Pi_4}{\Pi_5}, \frac{\Pi_1}{\Pi_2}, \frac{\Pi_4^2}{\Pi_1}\right) = f\left(Nu, Pr, Re, \frac{\Pi_1}{\Pi_2}, \frac{\Pi_4^2}{\Pi_1}\right) = 0$$

The solution in terms of the heat transfer coefficient is

$$Nu = \kappa \times f\left(Pr, Re, \frac{\Pi_1}{\Pi_2}, \frac{\Pi_4^2}{\Pi_1}\right)$$

But, the fourth and fifth terms in these functions are generally considered negligible when analyzing such processes. Thus

$$Nu = \kappa \times f(Pr, Re)$$

The total number of experiments required to define the above function is

$$N_{Expts}^{Total} = 5^2 = 25$$

If we solve this example dimensionally, we obtain the functional solution:

$$h = \kappa \times f(K_{Ave}, K_{Pipe}, D, v, \mu, \rho, C_P, k)$$

The required number of experiments to define the dimensional function is

$$N_{\text{Expts}}^{\text{Total}} = 5^8 = 390,625$$

This example, again, confirms the benefits of developing a project using Dimensional Analysis compared to our normal process development procedures.

6.7 SUMMARY

We progressed from three fundamental dimensions to four fundamental dimensions in this chapter. The additional fundamental dimension increased the complexity of our Dimensional Analysis by adding a row to the Dimension Table and by increasing the variables in each example. We estimated the number of experiments required to establish the function for each example and compared that estimation to the number of experiments required to confirm a dimensional result. The outcome of that comparison was, in each case: develop projects using Dimensional Analysis whenever possible.

REFERENCES

1. H. Huntley, *Dimensional Analysis*, Dover Publications, Inc., New York, NY, 1967.

2. T. Szirtes, *Applied Dimensional Analysis and Modeling*, Butterworth-Heinemann, Burlington, MA, 2007.

3. M. Zlokarnik, *Scale-up in Chemical Engineering*, Second Edition, Wiley-VCH Verlag GmbH & Co., KGaA, Weinheim, Germany, 2006.

4. A. Porter, *The Method of Dimensions*, Second Edition, Methuen & Co. Ltd., London, UK, 1943, pp. 57–59.

5. W. McAdams, *Heat Transmission*, Third Edition, McGraw-Hill Book Company, Inc., New York, NY, 1954, pp. 202–281.

CHAPTER 7

Mass Transfer and Reaction Examples of Dimensional Analysis

7.1 INTRODUCTION

Those engineering disciplines concerned with fluid flow, such as aeronautical, civil, and mechanical, have used Dimensional Analysis to good effect. Their success is largely attributable to the fact that fluid flow requires only three fundamental dimensions and generates a limited number of dimensionless parameters. Thus, the algebra is amenable to hand calculation.

Mechanical and chemical engineers are both concerned about heat flow, either into or from a given mechanism or process. Working with heat flow, i.e., heat transfer, requires a fourth fundamental dimension, namely, temperature or thermal energy, which complicates the algebra of Dimensional Analysis. And, where complicated algebra occurs, mistakes arise. Dimensional Analysis involving four fundamental dimensions has been done many times by hand, but such efforts involve significant amounts of time and effort to obtain the first solution, then to check that solution for possible algebraic errors. Thus, the application of Dimensional Analysis to situations involving heat transfer is much smaller than for those situations involving fluid flow.

The situation is even more complicated for chemical engineers, who are concerned with chemical change and with producing chemical products at acceptable rates. Analyzing chemical processes requires a fifth fundamental dimension, that dimension being "amount of substance," which is moles in the SI system of units. Chemical change also involves fluid flow and heat transfer, either initiated by the chemical reaction itself or required by the chemical process. Thus, the algebra for Dimensional Analysis of chemical processes is daunting. Due to the algebraic complexity of the effort, chemical engineers have not utilized Dimensional Analysis to the extent that other engineering disciplines have utilized it.

The matrix formulation of Dimensional Analysis and the availability of free-for-use matrix calculators on the Internet resolve the algebraic issues for chemical engineers and provide a rapid method for determining the dimensionless parameters best describing a chemical process.

7.2 FIRST-ORDER, HOMOGENEOUS BATCH REACTION

Consider a first-order chemical reaction occurring adiabatically in a batch reactor. The volume of the reactor is the volume of solvent used during the reaction. The agitator of this batch reactor is so efficient that no concentration differences exist within the process fluid; hence, diffusion of reactant is unimportant. The question is: how does C_F/C_S, where C_F is the final concentration of reactant and C_S is the starting concentration of reactant, relate to the other parameters of the process?

The geometric variable is reaction volume, i.e., the volume of solvent in the reactor V [L^3]. The material variables are solvent heat capacity C_P [$L^2MT^{-2}\theta^{-1}$] and solvent heat conductivity k [$LMT^{-3}\theta^{-1}$]. The process variables, those variables concerned with the reaction, are C_S and C_F [$L^{-3}N$], reaction time Δt [T], heat of reaction per unit time and volume [$L^{-1}MT^{-2}$], where heat of reaction is [$L^2MT^{-2}N^{-1}$] and C_S is [$L^{-3}N$]— combining these two terms yields the dimension of [$L^{-1}MT^{-2}$]. The effective reaction rate constant is k_R [T^{-1}], defined as

$$k_R = k_S \, e^{-(E/RT_S)}$$

where E is the energy of activation for the reaction and R is the gas constant and their ratio has dimension [θ]. The temperature process variables are ΔK_{WF} [θ], which is the temperature difference between the reaction fluid and the reactor wall, and the starting temperature T_S [θ]. We determine all the physical properties for this reaction at T_S.

The Dimensional Table is

Variables		V	Δt	k_S	E/R	ΔK_{WF}	C_F	C_P	k	$C_S\Delta H_R$	K_S	C_S
Dimensions	L	3	0	0	0	0	-3	2	1	-1	0	-3
	M	0	0	0	0	0	0	1	1	1	0	0
	T	0	1	-1	0	0	0	-2	-3	-2	0	0
	θ	0	0	0	1	1	0	-1	-1	0	1	0
	N	0	0	0	0	0	1	0	0	0	0	1

and the Dimension matrix is

$$
\begin{bmatrix}
3 & 0 & 0 & 0 & 0 & -3 & 2 & 1 & -1 & 0 & -3 \\
0 & 0 & 0 & 0 & 0 & 0 & 1 & 1 & 1 & 0 & 0 \\
0 & 1 & -1 & 0 & 0 & 0 & -2 & -3 & -2 & 0 & 0 \\
0 & 0 & 0 & 1 & 1 & 0 & -1 & -1 & 0 & 1 & 0 \\
0 & 0 & 0 & 0 & 0 & 1 & 0 & 0 & 0 & 0 & 1
\end{bmatrix}
$$

The largest square matrix for this Dimension matrix is 5×5, which is:

$$
R = \begin{bmatrix}
2 & 1 & -1 & 0 & -3 \\
1 & 1 & 1 & 0 & 0 \\
-2 & -3 & -2 & 0 & 0 \\
-1 & -1 & 0 & 1 & 0 \\
0 & 0 & 0 & 0 & 1
\end{bmatrix}
$$

Its determinant is

$$
|R| = \begin{vmatrix}
2 & 1 & -1 & 0 & -3 \\
1 & 1 & 1 & 0 & 0 \\
-2 & -3 & -2 & 0 & 0 \\
-1 & -1 & 0 & 1 & 0 \\
0 & 0 & 0 & 0 & 1
\end{vmatrix} = 3
$$

Thus, the Rank of this Dimension matrix is 5. The number of dimensionless parameters is

$$N_P = N_{Var} - R = 11 - 5 = 6$$

The inverse of R is

$$R^{-1} = \begin{bmatrix} 2 & 1 & -1 & 0 & -3 \\ 1 & 1 & 1 & 0 & 0 \\ -2 & -3 & -2 & 0 & 0 \\ -1 & -1 & 0 & 1 & 0 \\ 0 & 0 & 0 & 0 & 1 \end{bmatrix}^{-1} = \begin{bmatrix} 0.33 & 1.66 & 0.66 & 0 & 1 \\ 0 & -2 & -1 & 0 & 0 \\ -0.33 & 1.33 & 0.33 & 0 & -1 \\ 0.33 & -0.33 & -0.33 & 1 & 0 \\ 0 & 0 & 0 & 0 & 1 \end{bmatrix}$$

and the Bulk matrix is

$$B = \begin{bmatrix} 3 & 0 & 0 & 0 & 0 & -3 \\ 0 & 0 & 0 & 0 & 0 & 0 \\ 0 & 1 & -1 & 0 & 0 & 0 \\ 0 & 0 & 0 & 1 & 1 & 0 \\ 0 & 0 & 0 & 0 & 0 & 1 \end{bmatrix}$$

Therefore, $-R^{-1} \times B$ is

$$-R^{-1} \times B = -\begin{bmatrix} 0.33 & 1.66 & 0.66 & 0 & 1 \\ 0 & -2 & -1 & 0 & 0 \\ -0.33 & 1.33 & 0.33 & 0 & -1 \\ 0.33 & -0.33 & -0.33 & 1 & 0 \\ 0 & 0 & 0 & 0 & 1 \end{bmatrix} \begin{bmatrix} 3 & 0 & 0 & 0 & 0 & -3 \\ 0 & 0 & 0 & 0 & 0 & 0 \\ 0 & 1 & -1 & 0 & 0 & 0 \\ 0 & 0 & 0 & 1 & 1 & 0 \\ 0 & 0 & 0 & 0 & 0 & 1 \end{bmatrix}$$

$$-R^{-1} \times B = \begin{bmatrix} -1 & -0.66 & 0.66 & 0 & 0 & 0 \\ 0 & 1 & -1 & 0 & 0 & 0 \\ 1 & -0.33 & 0.33 & 0 & 0 & 0 \\ -1 & 0.33 & -0.33 & -1 & -1 & 0 \\ 0 & 0 & 0 & 0 & 0 & -1 \end{bmatrix}$$

Calculating matrix $(-R^{-1} \times B)^T$ shows that all the variables are pertinent to our analysis, i.e., $(-R^{-1} \times B)^T$ contains no columns entirely of zero.

$$(-R^{-1} \times B)^T = \begin{bmatrix} -1 & -0.66 & 0.66 & 0 & 0 & 0 \\ 0 & 1 & -1 & 0 & 0 & 0 \\ 1 & -0.33 & 0.33 & 0 & 0 & 0 \\ -1 & 0.33 & -0.33 & -1 & -1 & 0 \\ 0 & 0 & 0 & 0 & 0 & -1 \end{bmatrix}^T$$

$$= \begin{bmatrix} -1 & 0 & 1 & -1 & 0 \\ -0.66 & 1 & -0.33 & 0.33 & 0 \\ 0.66 & -1 & 0.33 & -0.33 & 0 \\ 0 & 0 & 0 & -1 & 0 \\ 0 & 0 & 0 & -1 & 0 \\ 0 & 0 & 0 & 0 & -1 \end{bmatrix}$$

The Total matrix is

	Π_1	Π_2	Π_3	Π_4	Π_5	Π_6					
V	1	0	0	0	0	0	0	0	0	0	0
Δt	0	1	0	0	0	0	0	0	0	0	0
k_S	0	0	1	0	0	0	0	0	0	0	0
E/R	0	0	0	1	0	0	0	0	0	0	0
ΔK_{WF}	0	0	0	0	1	0	0	0	0	0	0
C_F	0	0	0	0	0	1	0	0	0	0	0
C_P	−1	−0.66	0.66	0	0	0	−0.33	−1.66	−0.66	0	1
k	0	1	−1	0	0	0	0	2	1	0	0
$C_S\Delta H_R$	1	−0.33	0.33	0	0	0	0.33	−1.33	−0.33	0	−1
K_S	−1	0.33	−0.33	−1	−1	0	−0.33	0.33	0.33	−1	1
C_S	0	0	0	0	0	−1	0	0	0	0	1

$T =$

The dimensionless parameters, reading down each Π_i columns of the Total matrix, are

$$\Pi_1 = \frac{V(C_S\Delta H_R)}{C_P K_S}, \quad \Pi_2 = \frac{K_S^{0.33}k\Delta t}{C_P^{0.66}(C_S\Delta H_R)^{0.33}}, \quad \Pi_3 = \frac{k_S C_P^{0.66}(C_S\Delta H_R)^{0.33}}{K_S^{0.33}k},$$

$$\Pi_4 = \frac{E/R}{K_S}, \quad \Pi_5 = \frac{\Delta K_{WF}}{K_S}, \quad \Pi_6 = \frac{C_F}{C_S}$$

Note that Π_1 is a variant of the Group III Damkohler number Da^{III}.

It is best to use dimensionless parameters with whole number indices on their variables. Doing so makes it easier to interpret their physical meaning. We can achieve whole number indices on variables by combining the members of a given set of dimensionless parameters. For example, multiplying Π_3 by Π_2 yields

$$\Pi_3\Pi_2 = \left(\frac{k_S C_P^{0.66}(C_S\Delta H_R)^{0.33}}{K_S^{0.33}k}\right)\left(\frac{K_S^{0.33}k\Delta t}{C_P^{0.66}(C_S\Delta H_R)^{0.33}}\right) = k_S\Delta t$$

We require six dimensionless parameters for our solution: the sixth one is Π_4/Π_5.

The solution set for this example is

$$f\left(\Pi_1, \Pi_3\Pi_2, \Pi_4, \Pi_5, \Pi_4/\Pi_5, \Pi_6\right) = 0$$

and, in terms of C_F/C_S, the solution is

$$\Pi_6 = \kappa f\left(\Pi_1, \Pi_3\Pi_2, \Pi_4, \Pi_5, \Pi_4/\Pi_5\right)$$

In most cases, Π_4, Π_5, and Π_4/Π_5 small magnitudes and can be ignored. Doing so yields the solution

$$\Pi_6 = \kappa f(\Pi_1, \Pi_3\Pi_2)$$

The total number of experiments required to define the above function is

$$N_{Expts}^{Total} = 5^2 = 25$$

If we solve this example dimensionally, we obtain the functional solution:

$$\frac{C_F}{C_S} = \kappa \times f\left(V, \Delta t, k_S, E/R, \Delta K_{WF}, K_S, C_P, k, C_S \Delta H_R\right)$$

and the required experiments to define the function increase to

$$N_{Expts}^{Total} = 5^9 = 1.95 \times 10^6$$

which dramatically shows the benefits of developing a project using Dimensional Analysis compared to our normal process development procedures.

Of course, we would never do two million experiments during a process development effort. First, for this example, we assumed perfect mixing in the reaction volume; thus, ΔK_{WF} is probably too small to measure, thereby allowing us to exclude it from our list of independent variables. Second, we would assume that C_P and k had little impact on the chemical reaction, unless we had synthesized dimethyl kaboom, in which case we would be highly interested in the thermal properties of the solvent used during its synthesis and commercial production. Finally, we would measure ΔH_R early in our process development effort to determine how much heat the reaction generated or consumed during product formation, but, unless it was unusually large either way, we would exclude it from our list of independent process variables. The above function then reduces to

$$\frac{C_F}{C_S} = \kappa \times f\left(V, \Delta t, k_R, E/R, K_S\right)$$

which contains five independent variables. The required number of experiments for this function then becomes

$$N_{Expts}^{Total} = 5^5 = 3125$$

which is still significantly more than if we developed the process using Dimensional Analysis.

7.3 FIRST-ORDER, HOMOGENEOUS BATCH REACTION WITH AGITATION

Let us revisit the above example, but assume imperfect mixing. In other words, we need to include agitation in our analysis of the reaction. Thus, we add those variables related to agitation: the power P required to operate the agitator, the shaft rotational speed ω, the propeller or turbine diameter D, solvent viscosity μ $[L^{-1}MT^{-1}]$, and the solvent density ρ. The required power is measured, not calculated. Calculated power contains so many unverifiable assumptions that it is of little use to chemical engineers.

The geometric variables are reaction volume V $[L^3]$, i.e., the volume of solvent in the reactor and the propeller or turbine diameter D [L]. The material variables are solvent viscosity μ $[L^{-1}MT^{-1}]$, solvent density ρ $[L^{-3}M]$, solvent heat capacity C_P $[L^2MT^{-2}\theta^{-1}]$, and solvent heat conductivity k $[LMT^{-3}\theta^{-1}]$. The process variables are starting concentration C_S and final concentration C_F $[L^{-3}N]$, reaction time Δt [T], and heat of reaction $C_S\Delta H_R$ $[L^{-1}MT^{-2}]$. The effective reaction rate constant is k_R $[T^{-1}]$, defined as

$$k_R = k_S\, e^{-(E/RT_S)}$$

where E is the energy of activation for the reaction and R is the gas constant and their ratio has dimension $[\theta]$. The temperature process variables are ΔK_{WF} $[\theta]$, which is the temperature difference between the reaction fluid and the reactor wall, and the starting temperature T_S $[\theta]$. We determine all the physical properties for this reaction at T_S. The agitation process variables are required power P $[L^2MT^{-3}]$ and shaft rotational speed ω $[T^{-1}]$.

The Dimensional Table is

Variables	V	D	Δt	k_S	ω	E/R	ΔK_{WF}	K_S	$C_S\Delta H_R$	μ	C_F	C_S	ρ	P	C_P	k
L	3	1	0	0	0	0	0	0	−1	−1	−3	−3	−3	2	2	1
M	0	0	0	0	0	0	0	0	1	1	0	0	1	1	1	1
T	0	0	1	−1	−1	0	0	0	−2	−1	0	0	0	−3	−2	−3
θ	0	0	0	0	0	1	1	1	0	0	0	0	0	0	−1	−1
N	0	0	0	0	0	0	0	0	0	0	1	1	0	0	0	0

(Dimensions)

and the Dimension matrix is

$$
\begin{bmatrix}
3 & 1 & 0 & 0 & 0 & 0 & 0 & 0 & -1 & -1 & -3 & -3 & -3 & 2 & 2 & 1 \\
0 & 0 & 0 & 0 & 0 & 0 & 0 & 0 & 1 & 1 & 0 & 0 & 1 & 1 & 1 & 1 \\
0 & 0 & -1 & -1 & -1 & 0 & 0 & 0 & -2 & -1 & 0 & 0 & 0 & -3 & -2 & -3 \\
0 & 0 & 0 & 0 & 0 & 1 & 1 & 1 & 0 & 0 & 0 & 0 & 0 & 0 & -1 & -1 \\
0 & 0 & 0 & 0 & 0 & 0 & 0 & 0 & 0 & 0 & 1 & 1 & 0 & 0 & 0 & 0
\end{bmatrix}
$$

The largest square matrix for this Dimension matrix is 5×5, which is:

$$R = \begin{bmatrix} -3 & -3 & 2 & 2 & 1 \\ 0 & 1 & 1 & 1 & 1 \\ 0 & 0 & -3 & -2 & -3 \\ 0 & 0 & 0 & -1 & -1 \\ 1 & 0 & 0 & 0 & 0 \end{bmatrix}$$

Its determinant is

$$|R| = \begin{vmatrix} -3 & -3 & 2 & 2 & 1 \\ 0 & 1 & 1 & 1 & 1 \\ 0 & 0 & -3 & -2 & -3 \\ 0 & 0 & 0 & -1 & -1 \\ 1 & 0 & 0 & 0 & 0 \end{vmatrix} = 8$$

Thus, the Rank of this Dimension matrix is 5. The number of dimensionless parameters is

$$N_P = N_{Var} - R = 16 - 5 = 11$$

The inverse of R is

$$R^{-1} = \begin{bmatrix} -3 & -3 & 2 & 2 & 1 \\ 0 & 1 & 1 & 1 & 1 \\ 0 & 0 & -3 & -2 & -3 \\ 0 & 0 & 0 & -1 & -1 \\ 1 & 0 & 0 & 0 & 0 \end{bmatrix}^{-1}$$

$$= \begin{bmatrix} 0 & 0 & 0 & 0 & 1 \\ -0.125 & 0.625 & 0.125 & 0.125 & -0.375 \\ 0.125 & 0.375 & -0.125 & 0.875 & 0.375 \\ 0.375 & 1.125 & 0.625 & -0.375 & 1.125 \\ -0.375 & -1.125 & -0.625 & -0.625 & -1.125 \end{bmatrix}$$

and the Bulk matrix is

$$B = \begin{bmatrix} 3 & 1 & 0 & 0 & 0 & 0 & 0 & 0 & -1 & -1 & -3 \\ 0 & 0 & 0 & 0 & 0 & 0 & 0 & 0 & 1 & 1 & 0 \\ 0 & 0 & 1 & -1 & -1 & 0 & 0 & 0 & -2 & -1 & 0 \\ 0 & 0 & 0 & 0 & 0 & 1 & 1 & 1 & 0 & 0 & 0 \\ 0 & 0 & 0 & 0 & 0 & 0 & 0 & 0 & 0 & 0 & 1 \end{bmatrix}$$

Therefore, $-R^{-1} \times B$ is

$$
\begin{bmatrix}
0 & 0 & 0 & 0 & 0 & 0 & 0 & 0 & 0 & 0 & -1 \\
0.375 & 0.125 & -0.125 & 0.125 & 0.125 & -0.125 & -0.125 & -0.125 & -0.5 & -0.625 & 0 \\
-0.375 & -0.125 & 0.125 & -0.125 & -0.125 & -0.875 & -0.875 & -0.875 & -0.5 & -0.375 & 0 \\
-1.125 & -0.375 & -0.625 & 0.625 & 0.625 & 0.375 & 0.375 & 0.375 & 0.5 & -0.125 & 0 \\
1.125 & 0.375 & 0.625 & -0.625 & -0.625 & 0.625 & 0.625 & 0.625 & -0.5 & 0.125 & 0
\end{bmatrix}
$$

Calculating matrix $(-R^{-1} \times B)^{\mathrm{T}}$ shows that all the variables are pertinent to our analysis, i.e., $(-R^{-1} \times B)^{\mathrm{T}}$ contains no columns entirely of zero, as presented below.

$$
(-R^{-1} \times B)^{\mathrm{T}} =
\begin{bmatrix}
0 & 0.375 & -0.375 & -1.125 & 1.125 \\
0 & 0.125 & -0.125 & -0.375 & 0.375 \\
0 & -0.125 & 0.125 & -0.625 & 0.625 \\
0 & 0.125 & -0.125 & 0.625 & -0.625 \\
0 & 0.125 & -0.125 & 0.625 & -0.625 \\
0 & -0.125 & -0.875 & 0.375 & 0.625 \\
0 & -0.125 & -0.875 & 0.375 & 0.625 \\
0 & -0.125 & -0.875 & 0.375 & 0.625 \\
0 & -0.5 & -0.5 & 0.5 & -0.5 \\
0 & -0.625 & -0.375 & -0.125 & 0.125 \\
-1 & 0 & 0 & 0 & 0
\end{bmatrix}
$$

We can now assemble the Total matrix, which is shown below. Note that the lower, right portion of the Total matrix is presented as a partitioned matrix, namely, $[R]^{-1}$, in order to save space. That partitioned matrix does not enter the calculation determining the dimensionless variables.

$T =$

	Π_1	Π_2	Π_3	Π_4	Π_5	Π_6	Π_7	Π_8	Π_9	Π_{10}	Π_{11}		
V	1	0	0	0	0	0	0	0	0	0	0		
D	0	1	0	0	0	0	0	0	0	0	0		
Δt	0	0	1	0	0	0	0	0	0	0	0		
k_S	0	0	0	1	0	0	0	0	0	0	0		
ω	0	0	0	0	1	0	0	0	0	0	0		
E/R	0	0	0	0	0	1	0	0	0	0	0		
ΔK_{WF}	0	0	0	0	0	0	1	0	0	0	0		
K_S	0	0	0	0	0	0	0	1	0	0	0		
$C_S \Delta H_R$	0	0	0	0	0	0	0	0	1	0	0		
μ	0	0	0	0	0	0	0	0	0	1	0		
C_F	0	0	0	0	0	0	0	0	0	0	1		
C_S	0	0	0	0	0	0	0	0	0	0	-1		
ρ	0.375	0.125	-0.125	0.125	0.125	-0.125	-0.125	-0.125	-0.5	-0.625	0		
P	-0.375	-0.125	0.125	-0.125	-0.125	-0.875	-0.875	-0.875	-0.5	-0.375	0	$[R]^{-1}$	
C_P	-1.125	-0.375	-0.625	0.625	0.625	0.375	0.375	0.375	0.5	-0.125	0		
k	1.125	0.375	0.625	-0.625	-0.625	0.625	0.625	0.625	-0.5	0.125	0		

The dimensionless variables, reading down the Π_i columns of the Total matrix, are

$$\Pi_1 = \frac{V \rho^{0.375} k^{1.125}}{P^{0.375} C_P^{1.125}} \qquad \Pi_2 = \frac{D \rho^{0.125} k^{0.375}}{P^{0.125} C_P^{0.375}} \qquad \Pi_3 = \frac{\Delta t\, P^{0.125} k^{0.625}}{\rho^{0.125} C_P^{0.625}}$$

$$\Pi_4 = \frac{k_S \rho^{0.125} C_P^{0.625}}{P^{0.125} k^{0.625}} \qquad \Pi_5 = \frac{\omega \rho^{0.125} C_P^{0.625}}{P^{0.125} k^{0.625}} \qquad \Pi_6 = \frac{(E/R) C_P^{0.375} k^{0.625}}{\rho^{0.125} P^{0.875}}$$

$$\Pi_7 = \frac{\Delta K_{WF} C_P^{0.375} k^{0.625}}{\rho^{0.125} P^{0.875}} \qquad \Pi_8 = \frac{K_S C_P^{0.375} k^{0.625}}{\rho^{0.125} P^{0.875}} \qquad \Pi_9 = \frac{(C_S \Delta H_R) C_P^{0.5}}{\rho^{0.5} P^{0.5} k^{0.5}}$$

$$\Pi_{10} = \frac{\mu k^{0.125}}{\rho^{0.625} P^{0.375} C_P^{0.125}} \qquad \Pi_{11} = \frac{C_F}{C_S}$$

Combining dimensionless parameters in order to remove fractional powers or indices gives

$$\frac{\Pi_1}{\Pi_2^3} = \frac{V}{D^3} \qquad \frac{\Pi_4}{\Pi_5} = \frac{k_S}{\omega} \qquad \frac{\Pi_{10}}{\Pi_2^2\Pi_4} = \frac{\mu}{\rho D^2 k_S}$$

$$\frac{\Pi_1\Pi_4\Pi_9}{\Pi_5\Pi_8} = \frac{(C_S\Delta H_R)Vk_S}{K_S C_P \omega} \qquad \frac{\Pi_2^2\Pi_4\Pi_9}{\Pi_8} = \frac{(C_S\Delta H_R)D^2 k_S}{kK_S} \qquad \Pi_3\Pi_4 = k_S\Delta t$$

$$\Pi_3\Pi_5 = \omega\Delta t \qquad \frac{\Pi_7}{\Pi_8} = \frac{\Delta K_{WF}}{K_S} \qquad \frac{\Pi_6}{\Pi_8} = \frac{E}{RK_S}$$

$$\frac{\Pi_{10}}{\Pi_5\Pi_2^2} = \frac{\mu}{\rho D^2 \omega} = \frac{\mu}{\rho D(D\omega)} \qquad \Pi_{11} = \frac{C_F}{C_S}$$

Some of the above dimensionless parameters are readily interpreted. For example, Π_1/Π_2^3 is the ratio of reaction volume to the cube of the propeller or turbine diameter, which gives an indication of mixing efficiency, i.e., the smaller the number, the more efficient the mixing. $\Pi_3\Pi_4$ gives the extent of reaction during reaction time Δt, we will consider it dimensionless time. Π_6/Π_8 is the Arrhenius number and Π_7/Π_8 gives the impact of K_S on ΔK_{WF}. $\Pi_3\Pi_5$ is the dimensionless total rotation during reaction time Δt. With rearrangement, $\Pi_{10}/\Pi_5\Pi_2^2$ becomes the inverse Reynolds number. Π_{11} indicates the extent of reaction. The remaining dimensionless numbers are less intuitive. Π_4/Π_5 is a variant of the Group I Damkohler number Da^I, which is the ratio of chemical reaction rate to bulk mass flow rate.[1] If we multiply $\Pi_{10}/\Pi_2^2\Pi_4$ by D_{Diff}/D_{Diff}, i.e., by 1, where D_{Diff} is molecular diffusion, we obtain

$$\frac{\Pi_{10}}{\Pi_4\Pi_2^2} \times \frac{D_{Diff}}{D_{Diff}} = \frac{\mu}{\rho D^2 k_S} \times \frac{D_{Diff}}{D_{Diff}} = \frac{D_{Diff}}{k_S D^2} \times \frac{\mu}{\rho D_{Diff}}$$

which is the inverse Group II Damkohler number times the Schmidt number, symbolically

$$\frac{\Pi_{10}}{\Pi_4\Pi_2^2} \times \frac{D_{Diff}}{D_{Diff}} = \frac{Sc}{Da^{II}}$$

The Group II Damkohler number is the ratio of chemical reaction rate to molecular diffusion rate. For this example, we will consider $\Pi_{10}/\Pi_4\Pi_2^2$ a variant of $1/Da^{II}$. The fact that we multiply $\Pi_{10}/\Pi_2^2\Pi_4$ by D_{Diff}/D_{Diff} suggests that we should have included D_{Diff} as a variable in our Dimensional Analysis of this example; however, D_{Diff} is generally not considered a variable when analyzing chemical reactions in stirred, batch reactors. We generally assume that solvent agitation is so efficient that we can ignore molecular diffusion. $\Pi_1\Pi_4\Pi_9/\Pi_5\Pi_8$ is the Group III Damkohler number Da^{III}, which is the ratio of heat liberated or consumed by the chemical reaction to the bulk transport of heat. And, $\Pi_2^2\Pi_4\Pi_9/\Pi_8$ is the Group IV Damkohler number Da^{IV} that describes the ratio of heat liberated or consumed by the chemical reaction to conductive heat transfer.

The solution for this example is

$$f\left(\frac{\Pi_1}{\Pi_2^3},\Pi_3\Pi_4,\Pi_3\Pi_5,\frac{\Pi_6}{\Pi_8},\frac{\Pi_7}{\Pi_8},\frac{\Pi_{10}}{\Pi_2^2\Pi_5},\frac{\Pi_4}{\Pi_5},\frac{\Pi_{10}}{\Pi_2^2\Pi_4},\frac{\Pi_1\Pi_4\Pi_9}{\Pi_5\Pi_8},\frac{\Pi_2^2\Pi_4\Pi_9}{\Pi_8},\Pi_{11}\right)=0$$

or, in terms of C_F/C_S, the solution is

$$\Pi_{11}=\kappa\times f\left(\frac{\Pi_1}{\Pi_2^3},\Pi_3\Pi_4,\Pi_3\Pi_5,\frac{\Pi_6}{\Pi_8},\frac{\Pi_7}{\Pi_8},\frac{\Pi_{10}}{\Pi_2^2\Pi_5},\frac{\Pi_4}{\Pi_5},\frac{\Pi_{10}}{\Pi_2^2\Pi_4},\frac{\Pi_1\Pi_4\Pi_9}{\Pi_5\Pi_8},\frac{\Pi_2^2\Pi_4\Pi_9}{\Pi_8}\right)$$

The total number of experiments required by this solution is

$$N_{Expts}^{Total}=5^{10}=9.7\times10^6$$

No organization can afford the time or money to perform nearly 10 million experiments during a process development effort. This observation is well understood by chemical engineers: it is why chemical engineers consider process development and scaleup to be "arts." When developing a process, chemical engineers have historically used their experience to choose the variables important to a given process development. That is why process development and scaleup involve so much "rework": incorrect variables are chosen initially.

In the past, chemical engineers did not use Dimensional Analysis for process development, although during scaleup the concepts of similitude are followed. This example makes it clear why chemical engineers do not use Dimensional Analysis during a process development effort: process development involves too many variables and too many unspecified "degrees of freedom," i.e., too many indices require specification. However, the matrix formulation of Dimensional Analysis and the free-for-use matrix calculators available on the Internet now permit chemical engineers to employ the full power of Dimensional Analysis. By employing Dimensional Analysis during the initial phase of a process development, the chemical engineer can identify the dimensionless parameters controlling the process and estimate which dimensionless parameters are significant and which are insignificant. The insignificant dimensionless parameters can be neglected, thereby reducing the number of experiments required during a process development effort. For example, preliminary calculation, i.e., estimation, of each dimensionless parameter in the above solution will most likely show that the important dimensionless parameters are Π_1/Π_2^3, $\Pi_3\Pi_4$, Π_6/Π_8, and Π_4/Π_5. If the reaction involves significant thermal or mass transfer, then Sc/Da^{II}, Da^{III}, and Da^{IV} become important dimensionless variables. However, assuming the reaction to be homogeneous and well mixed precludes diffusion effects. Therefore, the solution for the above example becomes

$$\Pi_{11} = \kappa \times f\left(\frac{\Pi_1}{\Pi_2^3}, \Pi_3\Pi_4, \frac{\Pi_6}{\Pi_8}, \frac{\Pi_4}{\Pi_5}\right)$$

This solution involves $N_{\text{Expts}}^{\text{Total}} = 5^4 = 625$ experiments which can be done by 1 person in 2.5 years or by 2 people in 1.25 years, assuming 250 work days per year and one experiment per day.

Note that we have greatly reduced the number of experiments required for this process development effort by using a rational, logical procedure, a procedure that can be documented and retrieved for later use. In the past, such information has not been well documented nor has it been readily retrievable since decisions of this kind have been made via experiential knowledge.

This example clearly shows the benefit of using Dimensional Analysis during a process development effort.

7.4 FIRST-ORDER, HOMOGENEOUS REACTION IN A PLUG FLOW REACTOR[2]

Let us reconsider the above example, except instead of a batch reactor we will use a plug flow, tubular reactor. The process is adiabatic. We will now be concerned with the fluid velocity through the reactor instead of the agitation of the fluid. We will also be concerned with the reactor's diameter D and its length L as well, but not its volume *per se*. The reaction remains the same: the effective reaction rate constant is k_R $[T^{-1}]$, defined as

$$k_R = k_S\, e^{-(E/RT_S)}$$

where E is the energy of activation for the reaction and R is the gas constant and their ratio has dimension $[\theta]$.

The geometric variables, those variables defining reactor volume, are reactor diameter D [L] and reactor length L [L]. The material variables are fluid viscosity μ $[L^{-1}MT^{-1}]$, fluid density ρ $[L^{-3}M]$, fluid heat capacity C_P $[L^2MT^{-2}\theta^{-1}]$, and fluid heat conductivity k $[LMT^{-3}\theta^{-1}]$. We must consider the molecular diffusivity of reactant D_{Diff} $[L^2T^{-1}]$, both axially and radially, since we assume no back mixing in the reactor. The process variables are reactant inlet concentration C_{In} and reactant outlet concentration C_{Out} $[L^{-3}N]$, heat of reaction $C_{In}\Delta H_R$ $[L^{-1}MT^{-2}]$, fluid velocity through the reactor v $[LT^{-1}]$, temperature difference between inlet and outlet fluid ΔK_{IO} $[\theta]$, and inlet fluid temperature K_{In} $[\theta]$. We determine all the physical properties for this reaction at K_{In}.

The Dimensional Table is

<table>
<tr><td colspan="2">Variables</td><td>L</td><td>D</td><td>v</td><td>D_{Diff}</td><td>k_S</td><td>E/R</td><td>ΔK_{IO}</td><td>K_{In}</td><td>$C_{In}\Delta H_R$</td><td>C_{Out}</td><td>C_{In}</td><td>ρ</td><td>μ</td><td>C_P</td><td>k</td></tr>
<tr><td rowspan="5">Dimensions</td><td>L</td><td>1</td><td>1</td><td>1</td><td>2</td><td>0</td><td>0</td><td>0</td><td>0</td><td>-1</td><td>-3</td><td>-3</td><td>-3</td><td>-1</td><td>2</td><td>1</td></tr>
<tr><td>M</td><td>0</td><td>0</td><td>0</td><td>0</td><td>0</td><td>0</td><td>0</td><td>0</td><td>1</td><td>0</td><td>0</td><td>1</td><td>1</td><td>1</td><td>1</td></tr>
<tr><td>T</td><td>0</td><td>0</td><td>-1</td><td>-1</td><td>-1</td><td>0</td><td>0</td><td>0</td><td>-2</td><td>0</td><td>0</td><td>0</td><td>-1</td><td>-2</td><td>-3</td></tr>
<tr><td>θ</td><td>0</td><td>0</td><td>0</td><td>0</td><td>0</td><td>1</td><td>1</td><td>1</td><td>0</td><td>0</td><td>0</td><td>0</td><td>0</td><td>-1</td><td>-1</td></tr>
<tr><td>N</td><td>0</td><td>0</td><td>0</td><td>0</td><td>0</td><td>0</td><td>0</td><td>0</td><td>0</td><td>1</td><td>1</td><td>0</td><td>0</td><td>0</td><td>0</td></tr>
</table>

and the Dimension matrix is

$$
\begin{bmatrix}
1 & 1 & 1 & 2 & 0 & 0 & 0 & 0 & -1 & -3 & -3 & -3 & -1 & 2 & 1 \\
0 & 0 & 0 & 0 & 0 & 0 & 0 & 0 & 1 & 0 & 0 & 1 & 1 & 1 & 1 \\
0 & 0 & -1 & -1 & -1 & 0 & 0 & 0 & -2 & 0 & 0 & 0 & -1 & -2 & -3 \\
0 & 0 & 0 & 0 & 0 & 1 & 1 & 1 & 0 & 0 & 0 & 0 & 0 & -1 & -1 \\
0 & 0 & 0 & 0 & 0 & 0 & 0 & 0 & 0 & 1 & 1 & 0 & 0 & 0 & 0
\end{bmatrix}
$$

The largest square matrix for this Dimension matrix is 5×5, which is:

$$R = \begin{bmatrix} -3 & -3 & -1 & 2 & 1 \\ 0 & 1 & 1 & 1 & 1 \\ 0 & 0 & -1 & -2 & -3 \\ 0 & 0 & 0 & -1 & -1 \\ 1 & 0 & 0 & 0 & 0 \end{bmatrix}$$

Its determinant is

$$|R| = \begin{vmatrix} -3 & -3 & -1 & 2 & 1 \\ 0 & 1 & 1 & 1 & 1 \\ 0 & 0 & -1 & -2 & -3 \\ 0 & 0 & 0 & -1 & -1 \\ 1 & 0 & 0 & 0 & 0 \end{vmatrix} = 3$$

Thus, the Rank of this Dimension matrix is 5. The number of dimensionless parameters is

$$N_P = N_{Var} - R = 15 - 5 = 10$$

The inverse of R is

$$R^{-1} = \begin{bmatrix} -3 & -3 & -1 & 2 & 1 \\ 0 & 1 & 1 & 1 & 1 \\ 0 & 0 & -1 & -2 & -3 \\ 0 & 0 & 0 & -1 & -1 \\ 1 & 0 & 0 & 0 & 0 \end{bmatrix}^{-1} = \begin{bmatrix} 0 & 0 & 0 & 0 & 1 \\ -0.33 & 0 & 0.33 & -1.33 & -1 \\ 0.33 & 1 & -0.33 & 2.33 & 1 \\ 0.33 & 1 & 0.66 & -0.66 & 1 \\ -0.33 & -1 & -0.66 & -0.33 & -1 \end{bmatrix}$$

and the Bulk matrix is

$$B = \begin{bmatrix} 1 & 1 & 1 & 2 & 0 & 0 & 0 & 0 & -1 & -3 \\ 0 & 0 & 0 & 0 & 0 & 0 & 0 & 0 & 0 & 1 \\ 0 & 0 & -1 & -1 & -1 & 0 & 0 & 0 & -2 & 0 \\ 0 & 0 & 0 & 0 & 0 & 1 & 1 & 1 & 0 & 0 \\ 0 & 0 & 0 & 0 & 0 & 0 & 0 & 0 & 0 & 1 \end{bmatrix}$$

Therefore, $-R^{-1} \times B$ is

$$
\begin{bmatrix}
0 & 0 & 0 & 0 & 0 & 0 & 0 & 0 & 0 & -1 \\
0.33 & 0.33 & 0.66 & 1 & 0.33 & 1.33 & 1.33 & 1.33 & 0.33 & 0 \\
-0.33 & -0.33 & -0.66 & -1 & -0.33 & -2.33 & -2.33 & -2.33 & -1.33 & 0 \\
-0.33 & -0.33 & 0.33 & 0 & 0.66 & 0.66 & 0.66 & 0.66 & 0.66 & 0 \\
0.33 & 0.33 & -0.33 & 0 & -0.66 & 0.33 & 0.33 & 0.33 & -0.66 & 0
\end{bmatrix}
$$

Calculating matrix $(-R^{-1} \times B)^{T}$ shows that all the variables are pertinent to our analysis, i.e., there are no columns entirely of zero in $(-R^{-1} \times B)^{T}$, as shown below.

$$
(-R^{-1} \times B)^{T} =
\begin{bmatrix}
0 & 0.33 & -0.33 & -0.33 & 0.33 \\
0 & 0.33 & -0.33 & -0.33 & 0.33 \\
0 & 0.66 & -0.66 & 0.33 & -0.33 \\
0 & 1 & -1 & 0 & 0 \\
0 & 0.33 & -0.33 & 0.66 & -0.66 \\
0 & 1.33 & -2.33 & 0.66 & 0.33 \\
0 & 1.33 & -2.33 & 0.66 & 0.33 \\
0 & 1.33 & -2.33 & 0.66 & 0.33 \\
0 & 0.33 & -1.33 & 0.66 & -0.66 \\
-1 & 0 & 0 & 0 & 0
\end{bmatrix}
$$

We can now assemble the Total matrix, which is shown below. Note that the lower, right portion of the Total matrix is presented as a partitioned matrix, namely, $[R]^{-1}$, in order to save space. That partitioned matrix does not enter the calculation determining the dimensionless variables. The dimensionless variables are, reading down the Π_i columns of the Total matrix, given below.

		Π_1	Π_2	Π_3	Π_4	Π_5	Π_6	Π_7	Π_8	Π_9	Π_{10}						
	L	1	0	0	0	0	0	0	0	0	0	0	0	0	0	0	0
	D	0	1	0	0	0	0	0	0	0	0	0	0	0	0	0	0
	v	0	0	1	0	0	0	0	0	0	0	0	0	0	0	0	0
	D_{Diff}	0	0	0	1	0	0	0	0	0	0	0	0	0	0	0	0
	k_S	0	0	0	0	1	0	0	0	0	0	0	0	0	0	0	0
	E/R	0	0	0	0	0	1	0	0	0	0	0	0	0	0	0	0
	ΔK_{IO}	0	0	0	0	0	0	1	0	0	0	0	0	0	0	0	0
$T =$	K_{In}	0	0	0	0	0	0	0	1	0	0	0	0	0	0	0	0
	$C_{In}\Delta H_R$	0	0	0	0	0	0	0	0	1	0	0	0	0	0	0	0
	C_{Out}	0	0	0	0	0	0	0	0	0	1	0	0	0	0	0	0
	C_{In}	0	0	0	0	0	0	0	0	0	−1						
	ρ	0.33	0.33	0.66	1	0.33	1.33	1.33	1.33	0.33	0						
	μ	−0.33	−0.33	−0.66	−1	−0.33	−2.33	−2.33	−2.33	−1.33	0	$[R]^{-1}$					
	C_P	−0.33	−0.33	0.33	0	0.66	0.66	0.66	0.66	0.66	0						
	k	0.33	0.33	−0.33	0	−0.66	0.33	0.33	0.33	−0.66	0						

$$\Pi_1 = \frac{L\rho^{0.33}k^{0.33}}{\mu^{0.33}C_P^{0.33}} \qquad \Pi_2 = \frac{D\rho^{0.33}k^{0.33}}{\mu^{0.33}C_P^{0.33}} \qquad \Pi_3 = \frac{v\rho^{0.66}C_P^{0.33}}{\mu^{0.66}k^{0.33}}$$

$$\Pi_4 = \frac{D_{Diff}\rho}{\mu} \qquad \Pi_5 = \frac{k_S\rho^{0.33}C_P^{0.66}}{\mu^{0.33}k^{0.66}} \qquad \Pi_6 = \frac{(E/R)\rho^{1.33}C_P^{0.66}}{\mu^{2.33}}$$

$$\Pi_7 = \frac{\Delta K_{IO}\rho^{1.33}C_P^{0.66}k^{0.33}}{\mu^{2.33}} \qquad \Pi_8 = \frac{K_{In}\rho^{1.33}C_P^{0.66}k^{0.33}}{\mu^{2.33}} \qquad \Pi_9 = \frac{(C_{In}\Delta H_R)\rho^{0.33}C_P^{0.66}}{\mu^{1.33}k^{0.66}}$$

$$\Pi_{10} = \frac{C_{Out}}{C_{In}}$$

Combining dimensionless parameters in order to remove fractional powers or indices gives

$$\frac{\Pi_1}{\Pi_2} = \frac{L}{D}$$

$$\Pi_2\Pi_3 = \frac{\rho D v}{\mu} = Re \qquad \Pi_4 = \frac{\rho D_{\text{Diff}}}{\mu} = Sc^{-1}$$

$$\frac{\Pi_1\Pi_5}{\Pi_3} = \frac{k_S L}{v} = \tau k_S = Da^{\text{I}} \qquad \frac{\Pi_1^2\Pi_5}{\Pi_4} = \frac{L^2 k_S}{D_{\text{Diff}}} = Da^{\text{II}} \qquad \frac{\Pi_1\Pi_5\Pi_9}{\Pi_3\Pi_8} = \frac{(C_{\text{In}}\Delta H_{\text{R}})L k_S \mu}{\rho v k K_{\text{In}}}$$

$$\frac{\Pi_1^2\Pi_5\Pi_9}{\Pi_8} = \frac{(C_{\text{In}}\Delta H_{\text{R}})L^2 k_S}{k K_{\text{In}}} = Da^{\text{IV}} \qquad \frac{\Pi_6}{\Pi_8} = \frac{E}{R K_{\text{In}}} \qquad \frac{\Pi_7}{\Pi_8} = \frac{\Delta K_{\text{IO}}}{K_{\text{In}}}$$

$$\Pi_{10} = \frac{C_{\text{Out}}}{C_{\text{In}}}$$

Most of the above dimensionless parameters are familiar to us. Π_1/Π_2 is the aspect ratio of the tubular reactor. $\Pi_2\Pi_3$ is the Reynolds number. Π_4 is the inverse Schmidt number. $\Pi_1\Pi_5/\Pi_3$ is the Group I Damkohler number, where τ is the average residence time for a reactant molecule in the reactor. $\Pi_1^2\Pi_5/\Pi_4$ is the Group II Damkohler number. Π_6/Π_8 is the Arrhenius dimensionless number and Π_7/Π_8 gives the impact of inlet fluid temperature on ΔK_{IO}. Π_{10} gives the efficiency of the reaction. If we multiply $\Pi_1\Pi_5\Pi_9/\Pi_3\Pi_8$ by $C_{\text{P}}/C_{\text{P}}$, i.e., by 1, we obtain

$$\frac{\Pi_1\Pi_5\Pi_9}{\Pi_3\Pi_8} \times \frac{C_{\text{P}}}{C_{\text{P}}} = \frac{(C_{\text{In}}\Delta H_{\text{R}})L k_S \mu}{\rho v k K_{\text{In}}} \times \frac{C_{\text{P}}}{C_{\text{P}}} = \frac{(C_{\text{In}}\Delta H_{\text{R}})L k_S}{\rho v C_{\text{P}} K_{\text{In}}} \times \frac{\mu C_{\text{P}}}{k}$$

$$= Da^{\text{III}} \times Pr$$

where Pr is the Prandtl number, which is the ratio of momentum diffusivity to thermal diffusivity.

The solution for this example is

$$f\left(\frac{\Pi_1}{\Pi_2}, \Pi_2\Pi_3, \Pi_4, \frac{\Pi_1\Pi_5}{\Pi_3}, \frac{\Pi_1^2\Pi_5}{\Pi_4}, \frac{\Pi_1\Pi_5\Pi_9}{\Pi_3\Pi_8}, \frac{\Pi_1^2\Pi_5\Pi_9}{\Pi_8}, \frac{\Pi_6}{\Pi_8}, \frac{\Pi_7}{\Pi_8}, \Pi_{10}\right) = 0$$

or, in terms of $C_{\text{F}}/C_{\text{S}}$, the solution is

$$\Pi_{10} = \kappa \times f\left(\frac{\Pi_1}{\Pi_2}, \Pi_2\Pi_3, \Pi_4, \frac{\Pi_1\Pi_5}{\Pi_3}, \frac{\Pi_1^2\Pi_5}{\Pi_4}, \frac{\Pi_1\Pi_5\Pi_9}{\Pi_3\Pi_8}, \frac{\Pi_1^2\Pi_5\Pi_9}{\Pi_8}, \frac{\Pi_6}{\Pi_8}, \frac{\Pi_7}{\Pi_8}\right)$$

This solution requires a total of $N_{\text{Expts}}^{\text{Total}} = 5^9 = 1.95 \times 10^6$ experiments. Two million experiments are far too many to perform during a process development effort. At 1 experiment per day, it would require 1 person 7800 years to complete the effort. If we assume we have 3 years to complete this process development effort, then, at 1 experiment per person per day, we would need 2600 people working full-time on it. Such a project would be equivalent to the Manhattan Project during the Second World War or to the Race to the Moon during the 1960s. A project of such magnitude is outside the purview of a private organization. We can reduce the number of required experiments by using Dimensional Analysis, then determining which dimensionless parameters are critical to the function. For the above set of dimensionless parameters, we can calculate Π_4 and assume Π_7/Π_8 to be negligibly small. If the heat of reaction is small, we can neglect $\text{Da}^{\text{III}} \times \text{Pr}$ and Da^{IV}. Da^{II} can be estimated using information from Da^{I}. Therefore, with some forethought, we can reduce the above solution to

$$\Pi_{10} = \kappa \times f\left(\frac{\Pi_1}{\Pi_2}, \Pi_2\Pi_3, \frac{\Pi_1\Pi_5}{\Pi_3}, \frac{\Pi_6}{\Pi_8}\right)$$

This solution involves $N_{\text{Expts}}^{\text{Total}} = 5^4 = 625$ experiments which can be done by 1 person in 2.5 years, by 2 people in 1.25 years, and by 3 people in 0.8 years. This process development effort could be undertaken by a private organization.

7.5 FIRST-ORDER REACTION USING A SOLID SUPPORTED CATALYST

7.5.1 Introduction

Catalyzed reactions are necessary in the chemical processing industry in order to achieve commercially viable production rates. Many such catalysts are supported on porous solids. The solid structure of such catalysts provides strength while the porous feature provides surface area. Reaction rate depends upon solid surface area: the more surface area, the more catalytic sites available for product formation.

The CPI generally fills empty cylindrical towers with solid supported catalysts, thereby creating "fixed-bed" reactors. Such reactors are insulated and generally operated adiabatically. If the feed is a gas, it enters the reactor through a top nozzle, then flows downward through the catalyst bed. If the feed is a liquid, it enters the reactor through a bottom nozzle, then flows upward through the catalyst bed.

In fixed-bed, catalyzed processes, reactant molecules must first diffuse through the stagnant film surrounding every catalyst particle, pellet, or extrudate. The thickness of this stagnant film, thus the time to cross it by diffusion, is governed by the bulk physical properties of the fluid flowing through the catalyst bed. After passing through this stagnant film, the reactant molecules diffuse along a catalyst pore filled with quiescent fluid. At some point in time and at some point along the pore, the reactant molecules encounter an active, catalytic site where they are converted to product. The product molecules then reverse the diffusion steps to enter the flowing bulk fluid and exit the reactor.

Dimensionally analyzing this process as a sequence of steps will generate a large number of dimensionless parameters and would not necessarily identify which step controls the rate of product formation. A better plan is to identify which step is rate controlling, then analyze it independent of the other steps in the combined, actual process. We will pursue this plan and consider each possible rate limiting step separately. We will also use first-order mechanisms to represent the diffusion controlling steps since diffusion can be modeled as concentration differences. We will assume the chemical reaction at the catalytic site is first order.

7.5.2 Stagnant Film Diffusion Rate Limited Catalysis

Consider a cylindrical tower filled with a porous solid support. The feed is liquid and flows upward through the reactor. The reactor operates adiabatically.

The geometric variables, the variables defining the volume of catalyst, are reactor diameter D [L] and catalyst bed length length L [L]. Note that these are actually catalyst bed length and catalyst bed diameter. The material variables are fluid viscosity μ [$L^{-1}MT^{-1}$], fluid density ρ [$L^{-3}M$], fluid–solid heat capacity C_P [$L^2MT^{-2}\theta^{-1}$], fluid–solid heat conductivity k [$LMT^{-3}\theta^{-1}$], and molecular diffusivity D_{Diff} [L^2T^{-1}]. The process variables are

- reactant concentration entering the reactor C_{In} and reactant concentration exiting the reactor C_{Out} [$L^{-3}N$];
- the heat of reaction $C_{In}\Delta H_R$ [$L^{-1}MT^{-2}$], where ΔH_R has dimensions of $L^2MT^{-2}N^{-1}$ and C_{In} has dimensions of [$L^{-3}N$];

- the interstitial fluid velocity through the reactor v $[LT^{-1}]$—interstitial velocity is $v = Q/\varepsilon A$, where Q is volumetric flow rate, A is the cross-sectional area of the empty cylindrical tower, and ε is the void fraction of the porous solid catalyst;
- the fluid temperature entering the reactor K_{In} $[\theta]$—we determine all physical properties at K_{In};
- the temperature difference between the entering fluid and exiting fluid ΔK_{IO} $[\theta]$;
- the stagnant film diffusion rate constant k_{SF} (S/V) $[T^{-1}]$, where k_{SF} has dimensions of LT^{-1} and S/V, which is the ratio of external surface area S $[L^2]$ to physical volume V $[L^3]$ of the catalyst particle, pellet, or extrudate, has dimensions of L^{-1}.

The Dimensional Table is

Variables		L	D	v	D_{Diff}	k_{SF} (S/V)	K_{In}	ΔK_{IO}	$C_{In}\Delta H_R$	C_{Out}	C_{In}	ρ	μ	C_P	k
Dimensions	L	1	1	1	2	0	0	0	-1	-3	-3	-3	-1	2	1
	M	0	0	0	0	0	0	0	1	0	0	1	1	1	1
	T	0	0	-1	-1	-1	0	0	-2	0	0	0	-1	-2	-3
	θ	0	0	0	0	0	1	1	0	0	0	0	0	-1	-1
	N	0	0	0	0	0	0	0	0	1	1	0	0	0	0

and the Dimension matrix is

$$\begin{bmatrix} 1 & 1 & 1 & 2 & 0 & 0 & 0 & -1 & -3 & -3 & -3 & -1 & 2 & 1 \\ 0 & 0 & 0 & 0 & 0 & 0 & 0 & 1 & 0 & 0 & 1 & 1 & 1 & 1 \\ 0 & 0 & -1 & -1 & -1 & 0 & 0 & -2 & 0 & 0 & 0 & -1 & -2 & -3 \\ 0 & 0 & 0 & 0 & 0 & 1 & 1 & 0 & 0 & 0 & 0 & 0 & -1 & -1 \\ 0 & 0 & 0 & 0 & 0 & 0 & 0 & 0 & 1 & 1 & 0 & 0 & 0 & 0 \end{bmatrix}$$

The largest square matrix for this Dimension matrix is 5×5, which is:

$$R = \begin{bmatrix} -3 & -3 & -1 & 2 & 1 \\ 0 & 1 & 1 & 1 & 1 \\ 0 & 0 & -1 & -2 & -3 \\ 0 & 0 & 0 & -1 & -1 \\ 1 & 0 & 0 & 0 & 0 \end{bmatrix}$$

Its determinant is

$$|R| = \begin{vmatrix} -3 & -3 & -1 & 2 & 1 \\ 0 & 1 & 1 & 1 & 1 \\ 0 & 0 & -1 & -2 & -3 \\ 0 & 0 & 0 & -1 & -1 \\ 1 & 0 & 0 & 0 & 0 \end{vmatrix} = 3$$

Since $|R|$ is 3, the Rank of this Dimension matrix is 5. The number of dimensionless parameters is

$$N_P = N_{Var} - R = 14 - 5 = 9$$

The inverse of R is

$$R^{-1} = \begin{bmatrix} -3 & -3 & -1 & 2 & 1 \\ 0 & 1 & 1 & 1 & 1 \\ 0 & 0 & -1 & -2 & -3 \\ 0 & 0 & 0 & -1 & -1 \\ 1 & 0 & 0 & 0 & 0 \end{bmatrix}^{-1} = \begin{bmatrix} 0 & 0 & 0 & 0 & 1 \\ -0.33 & 0 & 0.33 & -1.33 & -1 \\ 0.33 & 1 & -0.33 & 2.33 & 1 \\ 0.33 & 1 & 0.66 & -0.66 & 1 \\ -0.33 & -1 & -0.66 & -0.33 & -1 \end{bmatrix}$$

and the Bulk matrix is

$$B = \begin{bmatrix} 1 & 1 & 1 & 2 & 0 & 0 & 0 & -1 & -3 \\ 0 & 0 & 0 & 0 & 0 & 0 & 0 & 1 & 0 \\ 0 & 0 & -1 & -1 & -1 & 0 & 0 & -2 & 0 \\ 0 & 0 & 0 & 0 & 0 & 1 & 1 & 0 & 0 \\ 0 & 0 & 0 & 0 & 0 & 0 & 0 & 0 & 1 \end{bmatrix}$$

Therefore, $-R^{-1} \times B$ is

$$-R^{-1} \times B = \begin{bmatrix} 0 & 0 & 0 & 0 & 0 & 0 & 0 & 0 & -1 \\ 0.33 & 0.33 & 0.66 & 1 & 0.33 & 1.33 & 1.33 & 0.33 & 0 \\ -0.33 & -0.33 & -0.66 & -1 & -0.33 & -2.33 & -2.33 & -1.33 & 0 \\ -0.33 & -0.33 & 0.33 & 0 & 0.66 & 0.66 & 0.66 & 0.66 & 0 \\ 0.33 & 0.33 & -0.33 & 0 & -0.66 & 0.33 & 0.33 & -0.66 & 0 \end{bmatrix}$$

Calculating matrix $(-R^{-1} \times B)^T$ shows all the variables are pertinent to our analysis, i.e., there are no columns entirely of zero in $(-R^{-1} \times B)^T$, as shown below.

$$(-R^{-1} \times B)^{\mathrm{T}} = \begin{bmatrix} 0 & 0.33 & -0.33 & -0.33 & 0.33 \\ 0 & 0.33 & -0.33 & -0.33 & 0.33 \\ 0 & 0.66 & -0.66 & 0.33 & -0.33 \\ 0 & 1 & -1 & 0 & 0 \\ 0 & 0.33 & -0.33 & 0.66 & -0.66 \\ 0 & 1.33 & -2.33 & 0.66 & 0.33 \\ 0 & 1.33 & -2.33 & 0.66 & 0.33 \\ 0 & 1.33 & -2.33 & 0.66 & 0.33 \\ 0 & 0.33 & -1.33 & 0.66 & -0.66 \\ -1 & 0 & 0 & 0 & 0 \end{bmatrix}$$

We can now assemble the Total matrix, which is shown below. Note that the Total matrix contains two partitioned matrices, the zero Total matrix matrix [0] and the inverse of the Rank matrix $[R]^{-1}$. These partitioned matrices do not enter the calculation determining the dimensionless variables. The dimensionless parameters, reading down the Π_i columns of the Total matrix, are

$$T = \begin{array}{c} \\ L \\ D \\ v \\ D_{\mathrm{Diff}} \\ k_{\mathrm{SF}}(S/V) \\ K_{\mathrm{In}} \\ \Delta K_{\mathrm{IO}} \\ C_{\mathrm{In}}\Delta H_{\mathrm{R}} \\ C_{\mathrm{Out}} \\ C_{\mathrm{In}} \\ \rho \\ \mu \\ C_{\mathrm{P}} \\ k \end{array} \begin{array}{c} \Pi_1 \quad \Pi_2 \quad \Pi_3 \quad \Pi_4 \quad \Pi_5 \quad \Pi_6 \quad \Pi_7 \quad \Pi_8 \quad \Pi_9 \end{array}$$

	Π_1	Π_2	Π_3	Π_4	Π_5	Π_6	Π_7	Π_8	Π_9	
L	1	0	0	0	0	0	0	0	0	
D	0	1	0	0	0	0	0	0	0	
v	0	0	1	0	0	0	0	0	0	
D_{Diff}	0	0	0	1	0	0	0	0	0	[0]
$k_{\mathrm{SF}}(S/V)$	0	0	0	0	1	0	0	0	0	
K_{In}	0	0	0	0	0	1	0	0	0	
ΔK_{IO}	0	0	0	0	0	0	1	0	0	
$C_{\mathrm{In}}\Delta H_{\mathrm{R}}$	0	0	0	0	0	0	0	1	0	
C_{Out}	0	0	0	0	0	0	0	0	1	
C_{In}	0	0	0	0	0	0	0	0	-1	
ρ	0.33	0.33	0.66	1	0.33	1.33	1.33	0.33	0	
μ	-0.33	-0.33	-0.66	-1	-0.33	-2.33	-2.33	-1.33	0	$[R]^{-1}$
C_{P}	-0.33	-0.33	0.33	0	0.66	0.66	0.66	0.66	0	
k	0.33	0.33	-0.33	0	-0.66	0.33	0.33	-0.66	0	

$$\Pi_1 = \frac{L\rho^{0.33}k^{0.33}}{\mu^{0.33}C_{\mathrm{P}}^{0.33}} \qquad \Pi_2 = \frac{D\rho^{0.33}k^{0.33}}{\mu^{0.33}C_{\mathrm{P}}^{0.33}} \qquad \Pi_3 = \frac{v\rho^{0.66}C_{\mathrm{P}}^{0.33}}{\mu^{0.66}k^{0.33}}$$

$$\Pi_4 = \frac{D_{\mathrm{Diff}}\rho}{\mu} \qquad \Pi_5 = \frac{k_{\mathrm{SF}}(S/V)\rho^{0.33}C_{\mathrm{P}}^{0.66}}{\mu^{0.33}k^{0.66}} \qquad \Pi_6 = \frac{K_{\mathrm{In}}\rho^{1.33}C_{\mathrm{P}}^{0.66}}{\mu^{2.33}}$$

$$\Pi_7 = \frac{\Delta K_{\mathrm{IO}}\rho^{1.33}C_{\mathrm{P}}^{0.66}}{\mu^{2.33}} \qquad \Pi_8 = \frac{(C_{\mathrm{S}}\Delta H_{\mathrm{R}})\rho^{0.33}C_{\mathrm{P}}^{0.66}}{\mu^{1.33}k^{0.66}} \qquad \Pi_9 = \frac{C_{\mathrm{F}}}{C_{\mathrm{S}}}$$

Combining dimensionless parameters in order to remove fractional powers or indices gives

$$\frac{\Pi_1}{\Pi_2} = \frac{L}{D} \qquad\qquad \Pi_2\Pi_3 = \frac{\rho D v}{\mu} = \text{Re} \qquad\qquad \Pi_4 = \frac{\rho D_{\text{Diff}}}{\mu} = \text{Sc}^{-1}$$

$$\frac{\Pi_1\Pi_5}{\Pi_3} = \frac{Lk_{\text{SF}}(S/V)}{v} = \tau k_{\text{SF}}\left(\frac{S}{V}\right) \qquad \frac{\Pi_1^2\Pi_5}{\Pi_4} = \frac{L^2 k_{\text{SF}}(S/V)}{D_{\text{Diff}}} \qquad \frac{\Pi_1\Pi_5\Pi_8}{\Pi_3\Pi_6} = \frac{(C_{\text{In}}\Delta H_{\text{R}})Lk_{\text{SF}}(S/V)}{\rho k v K_{\text{In}}}$$

$$\frac{\Pi_1^2\Pi_5\Pi_8}{\Pi_6} = \frac{(C_{\text{In}}\Delta H_{\text{R}})L^2 k_{\text{SF}}(S/V)}{k K_{\text{In}}} \qquad \frac{\Pi_7}{\Pi_6} = \frac{\Delta K_{\text{IO}}}{K_{\text{In}}} \qquad\qquad \Pi_9 = \frac{C_{\text{Out}}}{C_{\text{In}}}$$

Π_1/Π_2 is the "aspect ratio" of the reactor, $\Pi_2\Pi_3$ is the Reynolds number, Π_4 is the inverse Schmidt number, which is the ratio of momentum diffusivity and molecular diffusivity. $\Pi_1\Pi_5/\Pi_3$ is average residence time for a reactant molecule to spend in the reactor; it is also the Group I Damkohler number Da^{I}, which is the ratio of chemical reaction rate to bulk flow rate, remembering that the chemical reaction rate is controlled by the diffusion rate of reactant across the stagnant film covering each solid catalyst surface in the reactor. $\Pi_1^2\Pi_5/\Pi_4$ is the Group II Damkohler number Da^{II}, which is the ratio of chemical reaction rate to molecular to diffusion rate. $\Pi_1^2\Pi_5\Pi_8/\Pi_6$ is the Group IV Damkohler number Da^{IV}, which is the ratio of heat liberated or consumed to conductive heat transfer. We must multiply $\Pi_1\Pi_5\Pi_8/\Pi_3\Pi_6$ by C_P/C_P, i.e., by 1, in order to obtain

$$\frac{(C_{\text{In}}\Delta H_{\text{R}})k_{\text{SF}}(S/V)L}{\rho v K_{\text{In}}C_P} \times \frac{\mu C_P}{k}$$

which is the Group III Damkohler number Da^{III} times the Prandtl number Pr. Da^{III} describes the ratio of heat liberated or consumed to the bulk transport of heat and Pr describes momentum diffusivity to thermal diffusivity. Π_7/Π_6 simply tells us how K_{In} impacts ΔK_{IO}. And, Π_9 provides information about the operating efficiency of the reactor.

The solution for this example is

$$f\left(\frac{\Pi_1}{\Pi_2}, \Pi_2\Pi_3, \Pi_4, \frac{\Pi_1\Pi_5}{\Pi_3}, \frac{\Pi_1^2\Pi_5}{\Pi_4}, \frac{\Pi_1\Pi_5\Pi_8}{\Pi_3\Pi_6}, \frac{\Pi_1^2\Pi_5\Pi_8}{\Pi_6}, \frac{\Pi_7}{\Pi_6}, \Pi_9\right) = 0$$

or, in terms of C_F/C_S, the solution is

$$\Pi_9 = \kappa \times f\left(\frac{\Pi_1}{\Pi_2}, \Pi_2\Pi_3, \Pi_4, \frac{\Pi_1\Pi_5}{\Pi_3}, \frac{\Pi_1^2\Pi_5}{\Pi_4}, \frac{\Pi_1\Pi_5\Pi_8}{\Pi_3\Pi_6}, \frac{\Pi_1^2\Pi_5\Pi_8}{\Pi_6}, \frac{\Pi_7}{\Pi_6}\right)$$

The total number of experiments required for this solution is $N_{\text{Expts}}^{\text{Total}} = 5^8 = 391,000$, which is a large number of experiments for a private organization to perform. The number of dimensionless parameters can be reduced by recognizing that Π_4 or Sc^{-1} is calculated from physical properties; thus, we can exclude it from our experiments "to be done" list. Also, $\Pi_1^2\Pi_5/\Pi_4$, $\Pi_1\Pi_5\Pi_8/\Pi_6$, and $\Pi_1^2\Pi_5\Pi_8/\Pi_3\Pi_6$, i.e., Da^{II}, Da^{III}, and Da^{IV}, respectively, can be estimated by calculation and the use of Da^I or $\Pi_1\Pi_5/\Pi_3$ since

$$\frac{Da^{II}}{Da^I \times Re} = Sc \text{ and } \frac{Da^{IV}}{Da^{III} \times Re} = Pr$$

Also, Π_7/Π_6 will be negligibly small; thus, it can be excluded from our experimental list. With these assumptions, our solution becomes

$$\Pi_9 = \kappa \times f\left(\frac{\Pi_1}{\Pi_2}, \Pi_2\Pi_3, \frac{\Pi_1\Pi_5}{\Pi_3}\right)$$

and the number of experiments required is $N_{\text{Expts}}^{\text{Total}} = 5^3 = 125$. A private organization can easily do this number of experiments during a process development effort.

Of course, if $C_{\text{In}}\Delta H_R$ is large, then we will need to perform many more experiments in order to quantify Da^{III} and Da^{IV}.

7.5.3 Pore Diffusion Rate Limited Catalysis

Now consider the same fixed-bed reactor, but assume that the process is pore diffusion rate limited. In this case, nothing external to the catalyst particle, pellet, or extrudate influences the rate of product formation.

There are two sets of geometric variables in this example: those variables related to reactor geometry and those related to the porous solid catalyst. The former variables are reactor diameter D [L] and reactor length L [L]. Note that these are actually catalyst bed length and catalyst bed diameter.

The latter variable is porous solid catalyst size. The smaller the porous solid catalyst, the shorter the distance traveled by a reactant molecule and a product molecule. Thus, the time either molecule spends in a pore decreases as the radius or diameter of the catalyst solid decreases. We generally express this pertinent variable as the ratio of the external surface area S [L^2] to the physical volume V [L^3] of the catalyst solid. Hence, as S/V [L^{-1}] increases, the pore length decreases and molecules spend less time in the pores. The material variables are fluid viscosity μ [L^{-1}MT^{-1}], fluid density ρ [L^{-3}M], fluid–solid heat capacity C_P [L^{2}MT$^{-2}\theta^{-1}$], fluid–solid heat conductivity k [LMT$^{-3}\theta^{-1}$], and molecular diffusivity D_{Diff} [L^{2}T^{-1}]. The process variables are

- reactant concentration entering the reactor C_{In} and reactant concentration exiting the reactor C_{Out} [L^{-3}N]—remember these are the only concentrations we can measure in such a process;
- the interstitial fluid velocity through the reactor v [LT^{-1}]—interstitial velocity is $v = Q/\varepsilon A$, where Q is volumetric flow rate, A is the cross-sectional area of the empty cylindrical tower, and ε is the void fraction of the porous solid catalyst;
- the heat of reaction $C_{In}\Delta H_R$ [L^{-1}MT^{-2}], where ΔH_R has dimensions of L^{2}MT^{-2}N^{-1} and C_{In} has dimensions of [L^{-3}N];
- the fluid temperature entering the reactor K_{In} [θ]—we determine all physical properties at K_{In};
- the temperature difference between the entering fluid and exiting fluid ΔK_{IO} [θ];
- the pore diffusion rate constant k_{PD} (a_P/v_P) [T^{-1}], where k_{PD} has dimensions of LT^{-1} and a_P/v_P, which is the ratio of pore cross-sectional area a_P [L^2] to pore volume v_P [L^3] of the catalyst particle, pellet, or extrudate, has dimensions of L^{-1}.

The Dimensional Table is

Variables		L	D	S/V	v	D_{Diff}	k_{PD} (a_P/v_P)	K_{In}	ΔK_{IO}	$C_{In}\Delta H_R$	C_{Out}	C_{In}	ρ	μ	C_P	k
Dimensions	L	1	1	−1	1	2	0	0	0	−1	−3	−3	−3	−1	2	1
	M	0	0	0	0	0	0	0	0	1	0	0	1	1	1	1
	T	0	0	0	−1	−1	−1	0	0	−2	0	0	0	−1	−2	−3
	θ	0	0	0	0	0	0	1	1	0	0	0	0	0	−1	−1
	N	0	0	0	0	0	0	0	0	0	1	1	0	0	0	0

and the Dimension matrix is

$$
\begin{bmatrix}
1 & 1 & -1 & 1 & 2 & 0 & 0 & 0 & -1 & -3 & -3 & -3 & -1 & 2 & 1 \\
0 & 0 & 0 & 0 & 0 & 0 & 0 & 0 & 1 & 0 & 0 & 1 & 1 & 1 & 1 \\
0 & 0 & 0 & -1 & -1 & -1 & 0 & 0 & -2 & 0 & 0 & 0 & -1 & -2 & -3 \\
0 & 0 & 0 & 0 & 0 & 0 & 1 & 1 & 0 & 0 & 0 & 0 & 0 & -1 & -1 \\
0 & 0 & 0 & 0 & 0 & 0 & 0 & 0 & 0 & 1 & 1 & 0 & 0 & 0 & 0
\end{bmatrix}
$$

The largest square matrix for this Dimension matrix is 5×5, which is as follows:

$$R = \begin{bmatrix} -3 & -3 & -1 & 2 & 1 \\ 0 & 1 & 1 & 1 & 1 \\ 0 & 0 & -1 & -2 & -3 \\ 0 & 0 & 0 & -1 & -1 \\ 1 & 0 & 0 & 0 & 0 \end{bmatrix}$$

Its determinant is

$$|R| = \begin{vmatrix} -3 & -3 & -1 & 2 & 1 \\ 0 & 1 & 1 & 1 & 1 \\ 0 & 0 & -1 & -2 & -3 \\ 0 & 0 & 0 & -1 & -1 \\ 1 & 0 & 0 & 0 & 0 \end{vmatrix} = 3$$

Since $|R|$ is 3, the Rank of this Dimension matrix is 5. The number of dimensionless parameters is

$$N_P = N_{Var} - R = 15 - 5 = 10$$

The inverse of R is

$$R^{-1} = \begin{bmatrix} -3 & -3 & -1 & 2 & 1 \\ 0 & 1 & 1 & 1 & 1 \\ 0 & 0 & -1 & -2 & -3 \\ 0 & 0 & 0 & -1 & -1 \\ 1 & 0 & 0 & 0 & 0 \end{bmatrix}^{-1} = \begin{bmatrix} 0 & 0 & 0 & 0 & 1 \\ -0.33 & 0 & 0.33 & -1.33 & -1 \\ 0.33 & 1 & -0.33 & 2.33 & 1 \\ 0.33 & 1 & 0.66 & -0.66 & 1 \\ -0.33 & -1 & -0.66 & -0.33 & -1 \end{bmatrix}$$

and the Bulk matrix is

$$B = \begin{bmatrix} 1 & 1 & -1 & 1 & 2 & 0 & 0 & 0 & -1 & -3 \\ 0 & 0 & 0 & 0 & 0 & 0 & 0 & 0 & 1 & 0 \\ 0 & 0 & 0 & -1 & -1 & -1 & 0 & 0 & -2 & 0 \\ 0 & 0 & 0 & 0 & 0 & 0 & 1 & 1 & 0 & 0 \\ 0 & 0 & 0 & 0 & 0 & 0 & 0 & 0 & 0 & 1 \end{bmatrix}$$

Therefore, $-R^{-1} \times B$ is

$$\begin{bmatrix} 0 & 0 & 0 & 0 & 0 & 0 & 0 & 0 & 0 & -1 \\ 0.33 & 0.33 & -0.33 & 0.66 & 1 & 0.33 & 1.33 & 1.33 & 0.33 & 0 \\ -0.33 & -0.33 & 0.33 & -0.66 & -1 & -0.33 & -2.33 & -2.33 & -1.33 & 0 \\ -0.33 & -0.33 & 0.33 & 0.33 & 0 & 0.66 & 0.66 & 0.66 & 0.66 & 0 \\ 0.33 & 0.33 & -0.33 & -0.33 & 0 & -0.66 & 0.33 & 0.33 & -0.66 & 0 \end{bmatrix}$$

Calculating matrix $(-R^{-1} \times B)^{\mathrm{T}}$ shows all the variables are pertinent to our analysis, i.e., there are no columns entirely of zero in $(-R^{-1} \times B)^{\mathrm{T}}$, as shown below.

$$(-R^{-1} \times B)^{\mathrm{T}} = \begin{bmatrix} 0 & 0.33 & -0.33 & -0.33 & 0.33 \\ 0 & 0.33 & -0.33 & -0.33 & 0.33 \\ 0 & -0.33 & 0.33 & 0.33 & -0.33 \\ 0 & 0.66 & -0.66 & 0.33 & -0.33 \\ 0 & 1 & -1 & 0 & 0 \\ 0 & 0.33 & -0.33 & 0.66 & -0.66 \\ 0 & 1.33 & -2.33 & 0.66 & 0.33 \\ 0 & 1..33 & -2.33 & 0.66 & 0.33 \\ 0 & 0.33 & -1.33 & 0.66 & -0.66 \\ -1 & 0 & 0 & 0 & 0 \end{bmatrix}$$

We can now assemble the Total matrix, noting that it contains two partitioned matrices, the zero matrix [0] and the inverse of the Rank matrix $[R]^{-1}$. These partitioned matrices do not enter the calculation determining the dimensionless variables.

	Π_1	Π_2	Π_3	Π_4	Π_5	Π_6	Π_7	Π_8	Π_9	Π_{10}	
L	1	0	0	0	0	0	0	0	0	0	
D	0	1	0	0	0	0	0	0	0	0	
S/V	0	0	1	0	0	0	0	0	0	0	
v	0	0	0	1	0	0	0	0	0	0	
D_{Diff}	0	0	0	0	1	0	0	0	0	0	
$k_{\mathrm{PD}}(a_{\mathrm{P}}/v_{\mathrm{P}})$	0	0	0	0	0	1	0	0	0	0	[0]
K_{In}	0	0	0	0	0	0	1	0	0	0	
$T =$ ΔK_{IO}	0	0	0	0	0	0	0	1	0	0	
$C_{\mathrm{In}}\Delta H_{\mathrm{R}}$	0	0	0	0	0	0	0	0	1	0	
C_{out}	0	0	0	0	0	0	0	0	0	1	
C_{In}	0	0	0	0	0	0	0	0	0	-1	
ρ	0.33	0.33	-0.33	0.66	1	0.33	1.33	1.33	0.33	0	
μ	-0.33	-0.33	0.33	-0.66	-1	-0.33	-2.33	-2.33	-1.33	0	$[R]^{-1}$
C_{P}	-0.33	-0.33	0.33	0.33	0	0.66	0.66	0.66	0.66	0	
k	0.33	0.33	-0.33	-0.33	0	-0.66	0.33	0.33	-0.66	0	

The dimensionless parameters, reading down the Π_i columns of the Total matrix, are

$$\Pi_1 = \frac{L\rho^{0.33}k^{0.33}}{\mu^{0.33}C_P^{0.33}} \qquad \Pi_2 = \frac{D\rho^{0.33}k^{0.33}}{\mu^{0.33}C_P^{0.33}} \qquad \Pi_3 = \frac{(S/V)\rho^{0.33}k^{0.33}}{\mu^{0.33}C_P^{0.33}}$$

$$\Pi_4 = \frac{v\rho^{0.66}C_P^{0.33}}{\mu^{0.66}k^{0.33}} \qquad \Pi_5 = \frac{D_{\text{Diff}}\rho}{\mu} = \text{Sc}^{-1} \qquad \Pi_6 = \frac{k_{PD}(a_P/v_P)\rho^{0.33}C_P^{0.66}}{\mu^{0.33}k^{0.66}}$$

$$\Pi_7 = \frac{K_{\text{In}}\rho^{1.33}C_P^{0.66}k^{0.33}}{\mu^{2.33}} \qquad \Pi_8 = \frac{\Delta K_{\text{IO}}\rho^{1.33}C_P^{0.66}k^{0.33}}{\mu^{2.33}} \qquad \Pi_9 = \frac{(C_S\Delta H_R)\rho^{0.33}C_P^{0.66}}{\mu^{1.33}k^{0.66}}$$

$$\Pi_{10} = \frac{C_{\text{out}}}{C_{\text{In}}}$$

Combining dimensionless parameters in order to remove fractional powers or indices gives

$$\frac{\Pi_1}{\Pi_2} = \frac{L}{D} \qquad\qquad \Pi_2\Pi_3 = \left(\frac{S}{V}\right)D \qquad\qquad \Pi_2\Pi_4 = \frac{\rho Dv}{\mu} = \text{Re}$$

$$\Pi_4 = \frac{\rho D_{\text{Diff}}}{\mu} = \text{Sc}^{-1} \qquad \frac{\Pi_1\Pi_6}{\Pi_4} = \frac{Lk_{PD}(a_P/v_P)}{v} = \tau k_{PD}\left(\frac{a_P}{v_P}\right) = Da^{\text{I}} \qquad \frac{\Pi_1^2\Pi_6}{\Pi_5} = \frac{L^2 k_{PD}(a_P/v_P)}{D_{\text{Diff}}} = Da^{\text{II}}$$

$$\frac{\Pi_1\Pi_6\Pi_9}{\Pi_4\Pi_7} = \frac{(C_{\text{In}}\Delta H_R)Lk_{PD}(a_P/v_P)}{\rho k v K_{\text{In}}} \qquad \frac{\Pi_1^2\Pi_6\Pi_9}{\Pi_7} = \frac{(C_{\text{In}}\Delta H_R)L^2 k_{PD}(a_P/v_P)}{k K_{\text{In}}} = Da^{\text{IV}} \qquad \frac{\Pi_8}{\Pi_7} = \frac{\Delta K_{\text{IO}}}{K_{\text{In}}}$$

$$\Pi_{10} = \frac{C_{\text{Out}}}{C_{\text{In}}}$$

We must multiply $\Pi_1\Pi_6\Pi_9/\Pi_4\Pi_7$ by C_P/C_P, i.e., by 1, in order to obtain

$$\frac{(C_{\text{In}}\Delta H_R)k_{PD}(a_P/v_P)L}{\rho v K_{\text{In}}C_P} \times \frac{\mu C_P}{k} = Da^{\text{III}} \times \text{Pr}$$

which is the Group III Damkohler number Da^{III} times the Prandtl number Pr. Da^{III} describes the ratio of heat liberated or consumed to the bulk transport of heat and Pr describes momentum diffusivity to thermal diffusivity.

The solution for this example is

$$f\left(\frac{\Pi_1}{\Pi_2}, \Pi_2\Pi_3, \Pi_2\Pi_4, \Pi_4, \frac{\Pi_1\Pi_6}{\Pi_4}, \frac{\Pi_1^2\Pi_6}{\Pi_5}, \frac{\Pi_1\Pi_6\Pi_9}{\Pi_4\Pi_7}, \frac{\Pi_1^2\Pi_6\Pi_9}{\Pi_7}, \frac{\Pi_8}{\Pi_7}, \Pi_{10}\right) = 0$$

or, in terms of C_F/C_S, the solution is

$$\Pi_9 = \kappa \times f\left(\frac{\Pi_1}{\Pi_2}, \Pi_2\Pi_3, \Pi_2\Pi_4, \Pi_4, \frac{\Pi_1\Pi_6}{\Pi_4}, \frac{\Pi_1^2\Pi_6}{\Pi_5}, \frac{\Pi_1\Pi_6\Pi_9}{\Pi_4\Pi_7}, \frac{\Pi_1^2\Pi_6\Pi_9}{\Pi_7}, \frac{\Pi_8}{\Pi_7}\right)$$

The total number of experiments required for this solution is $N_{\text{Expts}}^{\text{Total}} = 5^9 = 1.95 \times 10^6$, which is far too many experiments for a private organization to perform. The number of dimensionless parameters can be reduced by recognizing that Π_4 or Sc^{-1} is calculated from physical properties; thus, we can exclude it from our experiments "to be done" list. Also, $\Pi_1^2\Pi_5/\Pi_4$, $\Pi_1\Pi_5\Pi_8/\Pi_6$, and $\Pi_1^2\Pi_5\Pi_8/\Pi_3\Pi_6$, i.e., Da^{II}, Da^{III}, and Da^{IV}, respectively, can be estimated by calculation and the use of Da^{I} or $\Pi_1\Pi_5/\Pi_3$ since

$$\frac{\text{Da}^{\text{II}}}{\text{Da}^{\text{I}} \times \text{Re}} = \text{Sc} \quad \text{and} \quad \frac{\text{Da}^{\text{IV}}}{\text{Da}^{\text{III}} \times \text{Re}} = \text{Pr}$$

Also, Π_7/Π_6 will be negligibly small; thus, it can be excluded from our experimental list. With these assumptions, our solution becomes

$$\Pi_9 = \kappa \times f\left(\frac{\Pi_1}{\Pi_2}, \Pi_2\Pi_3, \Pi_2\Pi_4, \frac{\Pi_1\Pi_6}{\Pi_4}\right)$$

and the number of experiments required is $N_{\text{Expts}}^{\text{Total}} = 5^4 = 625$. A private organization can easily do this number of experiments during a process development effort.

Of course, if $C_{\text{In}}\Delta H_R$ is large, then we will need to perform many more experiments in order to quantify Da^{III} and Da^{IV}.

7.5.4 Reaction Rate Limited Catalysis

In this condition, neither diffusion across the stagnant film surrounding each catalyst particle, pellet, or extrudate nor diffusion within each porous solid adversely impacts product formation rate. The rate at which the reaction occurs at the catalytic site controls product formation rate.

The Dimensional Analysis of this catalytic condition is the same as for the stagnant film diffusion rate limited condition. The analysis, equations, and result are the same for these two catalytic conditions, except k_{Site}, the reaction rate at the catalytic site, replaces k_{SF} (S/V) in all the dimensionless parameters.

7.5.5 Summary

When developing a porous solid catalyzed process, we must first determine whether product formation rate is stagnant film diffusion rate limited, pore diffusion rate limited, or reaction rate limited. The experimental program to improve product formation rate depends upon the limiting condition. For stagnant film diffusion rate limited catalysis, reducing the thickness of the stagnant film surrounding each porous solid particle, pellet, or extrudate improves product formation rate. The simplest "fix" for this condition is to increase the Reynolds number for the process. Increasing S/V of the catalyst solid also improves product formation rate in stagnant film diffusion rate limited processes. With sufficient improvement, the process will shift from being stagnant film diffusion rate limited to being pore diffusion rate limited.

For pore diffusion rate limited catalysis, improving pore diffusion, i.e., increasing median pore diameter, increases product formation rate. Increasing the S/V ratio as of the solid catalyst also increases product formation rate since it reduces average pore length, which decreases the residence time molecules spent in catalyst pores. Again, with sufficient improvement, the process will shift from being pore diffusion rate limited to being reaction rate limited.

For reaction rate limited catalysis, increasing solid surface area increases the number of catalyst sites which increases product formation rate. Reducing median pore diameter increases surface area with the porous solid catalyst. Also, changing the chemistry of the catalytic site may increase product formation. For example, changing the valency of the catalytic metal, changing the ligands on the catalytic metal, or changing the catalytic metal may increase production formation rate.

But, to implement any of the above changes, we first must know which condition is limiting product formation rate. Plotting C_{Out}/C_{In} against Reynolds number with S/V ratio as the parameter will give insight as to which condition limits product formation. If the parametric line increases with increasing Reynolds number, then product formation is stagnant film diffusion rate limited. If the parametric line is independent of increasing Reynolds number, then product formation rate is pore diffusion rate limited. And, if the parametric line decreases with increasing Reynolds number, then reaction rate controls product formation rate. Note that each condition merges with another condition, so that sharp demarcations between conditions are rare. Generally, we see a gradual shift from one condition to another condition, thereby making it difficult to distinguish conditions.

7.6 SUMMARY

Chemical engineers have not used Dimensional Analysis to the extent that aeronautical, civil, or mechanical engineers have used it. This reluctance by chemical engineers to use Dimensional Analysis is due to the number of variables involved in a chemical process and to the number of fundamental dimensions quantifying each variable. Using Rayleigh's Method of Indices produces a horrific algebraic tangle which induces many mistakes when manipulating the resulting equations. Nondimensionalizing the three conservation laws underlying every chemical process does not provide much insight due to the impact each conservation law has upon another.

In this chapter, we applied the matrix formulation of Dimensional Analysis to chemical reactions occurring in batch reactors, plug flow reactors, and fixed-bed reactors. In each example, we generated a plethora of dimensionless parameters and showed how to reduce them to commonly used dimensionless parameters which have acquired names over time. The intricate algebra involved in obtaining these named dimensionless numbers is done by matrix manipulation, which is easily done using one of the free-for-use matrix calculators available on the Internet.

REFERENCES

1. D. Boucher, G. Alves, *Chemical Engineering Progress*, 55 (9), 55 (1959).
2. M. Zlokarnik, *Scale-up in Chemical Engineering*, Second Edition, Wiley-VCH Verlag GmbH & Co., KGaA, Weinheim, Germany, 2006, pp. 212–214.

CHAPTER 8

Dimensional Analysis and Scaling

8.1 INTRODUCTION

Process development, process support, and scaling go hand in hand: we cannot do process development without considering upscaling since moving the process to the commercial scale is the purpose of the effort. Likewise, we cannot do process support without considering downscaling. To support a commercial process, we have to model the troubled portion of the production plant in the laboratory or in a pilot plant. We cannot solve the production problem in the laboratory or in the pilot plant without considering upscaling the solution into the commercial plant. In other words, process development, process support, and scaling are, essentially, one and the same activity.

8.2 MODELING

Upscaling and downscaling involve modeling. We build models for the same reasons we use Dimensional Analysis during process development or process support: to reduce the time from idea to commercialization and to reduce the cost of the effort. The major cost savings of modeling come from not building an inoperable, full-scale commercial plant.

There are four types of models. They are

1. true models;
2. adequate models;
3. distorted models;
4. dissimilar models.

True models involve building all significant process features to scale. Thus, the model is an exact replica of the commercial plant, which we call the "prototype." We build true models in some safety investigations in order to determine definitely the cause of a grievous, horrific event. While true models may provide highly accurate information, they are capital intensive, expensive to operate, and require extended time periods to build.

Adequate models predict one characteristic of the prototype accurately. If the sizes of the model and the prototype are significantly different, then it is unlikely that we can achieve complete similarity. And, for complex processes, a complete model is actually a full-scale prototype, i.e., a true model.[1] If the modeled characteristic is the dominant, controlling factor in the process, then an adequate model may be sufficient. For example, porous solid catalyzed processes are generally stagnant film diffusion rate limited or pore diffusion rate limited. If a process is so limited, then we only have to ensure the same controlling regime in our laboratory or pilot plant reactors. If we do not ensure equivalent controlling regimes in the laboratory or pilot plant reactors, then any process development or process support will be wasted effort. If we do not consider whether the commercial process is stagnant film diffusion rate limited, pore diffusion rate limited, or reaction rate limited, then we will finish our effort with an expensive scattergram of the experimental results. This situation actually occurs more often than we like to admit. Many porous solid catalyzed commercial processes are pore diffusion rate limited due to high interstitial fluid velocity through the reactor. Such high fluid velocity minimizes the boundary layer surrounding each catalyst pellet, thereby making the process pore diffusion rate limited. Unfortunately, most porous solid catalyzed pilot plant processes, i.e., models, are operated at low interstitial fluid velocities in order to minimize feed and product volumes at the research site. In such situations, stagnant film diffusion rate is the controlling regime. The result of a multiyear, multicatalyst testing effort will be an expensive scattergram around the average value for the film diffusion rate constant. On the other hand, considerable effort can be made at the laboratory scale to ensure that catalyst testing occurs in the reaction rate limited regime. Plots with impressive correlations result from these types of experimental programs. Unfortunately, when the best catalyst is tested in the prototype, it displays the same efficiency and productivity as the current catalyst. In such cases, the prototype is either stagnant film or pore diffusion rate limited. It does not matter how reactive the catalyst is in the laboratory; in the prototype, the process is incapable of keeping the catalytic site saturated with reactant. In conclusion, the controlling regime of the model must be identical to the controlling regime of the prototype. With regard to the process, adequate models behave similarly to their prototypes, even though they may be many times smaller than their prototypes.

In distorted models, we violate design conditions intentionally for one reason or another. Such distortion affects the prediction equation. In other words, we have to correct data from the model in order to simulate the prototype. Hydrologic river basin models are the most common distorted models. In these models, the horizontal and vertical lengths do not have the same ratios or "scaling factors." In a geometrically similar model, the horizontal and vertical ratios are equal; for example

$$\frac{_pL_H}{_ML_H} = \kappa \quad \text{and} \quad \frac{_pL_V}{_ML_V} = \kappa$$

where $_pL_H$ is the prototype horizontal length of interest; $_ML_H$ is the model horizontal length equivalent to $_pL_H$; $_pL_V$ is the prototype vertical length of interest; and, $_ML_V$ is the model vertical length equivalent to $_pL_V$. κ is a constant, the "scaling factor." For a distorted model

$$\frac{_pL_H}{_ML_H} = \kappa \quad \text{and} \quad \frac{_pL_V}{_ML_V} = \lambda$$

where $\kappa \neq \lambda$. It is "legal" to use distorted models, just so long as we know we are doing it and we understand why we are doing it. With regard to the process, distorted models behave in a manner similar to their prototypes; however, one dimension of the model will not be scaled equivalently to the other dimensions. Thus, a distorted model may look squat or tall or broad, depending on the distortion, when compared to its prototype.[2]

Dissimilar models comprise the fourth and last model type. Such models have no apparent resemblance to the prototype. Dissimilar models have, as their name states, no similarity to their prototypes. These models provide information about the prototype through suitable analogies.

8.3 SIMILARITY

We base our models on similarity. Four similarities are important to chemical engineers. They are

1. geometrical;
2. mechanical;

3. thermal;
4. chemical.

In general, geometric similarity means that given two objects of different size, if there is a point within the smaller object, which we identify as the model, with coordinates x_M, y_M, and z_M, and a similar point within the larger object, i.e., the prototype, with coordinates x_P, y_P, and z_P, then the two objects are similar at that given point if

$$\frac{x_P}{x_M} = \frac{y_P}{y_M} = \frac{z_P}{z_M} = L$$

The two objects are geometrically similar if the above condition holds for all corresponding points within the two objects.

Mechanical similarity comprises three subsimilarities, which are static similarity, kinematic similarity, and dynamic similarity. Static similarity demands that two geometrically similar objects have relative deformation for a constant applied stress. This similarity is of interest to civil and structural engineers.

Kinematic similarity means the constituent parts of a model and prototype mechanism or process in translation follow similar paths or streamlines if the model and prototype are geometrically similar. Thus

$$\frac{v_P}{v_M} = V$$

where v_M is the velocity of the translating model part or particle and v_P is the translating prototype part or particle. V is the velocity scaling factor.

Dynamic similarity demands the ratio of the forces inducing acceleration be equal at corresponding locations in geometrically similar mechanisms or processes. In other words, the ratio

$$\frac{F_P}{F_M} = F$$

where F_M is the force at location x_M, y_M, and z_M in the model and F_P is the force at location x_P, y_P, and z_P in the prototype, holds true at every corresponding location in the two mechanisms or processes.

Thermal similarity occurs when the ratio of the temperature difference at corresponding locations of a geometrically similar mechanism or process is equal. If translation, i.e., movement, occurs, then the process must also demonstrate kinematic similarity for thermal similarity to occur. Thus, thermal similarity requires geometric similarity and kinematic similarity.

As chemical engineers, our major concern is the reactions occurring in a process. We want our model to reflect what occurs in our prototype. To ensure that outcome, our model must be chemically similar to our prototype. Chemical similarity demands the ratio of concentration differences at all corresponding locations in the model and in the prototype be equal. Therefore, our model and prototype must also be geometrically, mechanically, and thermally similar.

Consider two mechanical processes involving the Navier–Stokes equation. Let one process be large and the other process be small. Our question: is the larger process similar to the smaller process? The best way to answer that question is to convert the Navier–Stokes equation into a dimensionless form. Let us define a characteristic length L and velocity V, then form the dimensionless variables x^* and v^*, which are

$$x_S^* = \frac{x_S}{L} \quad \text{and} \quad v_{x,S}^* = \frac{v_{x,S}}{V}$$

where the subscript S identifies the small process; x, the length in the x-direction; L, the characteristic length; x_S^*, the dimensionless length in the x-direction; $v_{x,S}$, the fluid velocity in the x-direction in the small process; V, the characteristic velocity; and $v_{x,S}^*$, the dimensionless velocity in the x-direction. We define dimensionless pressure as

$$p^* = \frac{p}{\rho V^2}$$

and we define dimensionless time as

$$t^* = \frac{Vt}{L}$$

The Navier–Stokes equation in one-dimension for the small process is

$$\frac{\partial v_{x,S}}{\partial t} + v_{x,S} \frac{\partial v_{x,S}}{\partial x_S} = -\frac{1}{\rho} \frac{\partial p_S}{\partial x_S} + g_x + \frac{\mu}{\rho} \frac{\partial^2 v_{x,S}}{\partial (x_S)^2}$$

where p_S is the pressure of the small process; g_x is the acceleration due to gravity; μ is the fluid dynamic viscosity; and ρ is the fluid density. $\frac{\partial v_{x,S}}{\partial t}$ represents the local acceleration of the fluid particle; $v_{x,S} \frac{\partial v_{x,S}}{\partial x_S}$ is the convective acceleration of the fluid particle; $\frac{1}{\rho}\frac{\partial p_S}{\partial x_S}$ represents the pressure acceleration due to pumping action; and $\frac{\mu}{\rho}\frac{\partial^2 v_{x,S}}{\partial (x_S)^2}$ is the viscous deceleration generated by objects in the fluid's flow path.[3] Converting the dimensional equation to a dimensionless equation yields

$$\left(\frac{V^2}{L}\right)\frac{\partial v_{x,S}^*}{\partial t^*} + \left(\frac{V^2}{L}\right)v_{x,S}^*\frac{\partial v_{x,S}^*}{\partial x_S^*} = -\left(\frac{V^2}{L}\right)\frac{\partial p_S^*}{\partial x_S^*} + g_x + \frac{\mu V}{\rho L^2}\frac{\partial^2 v_{x,S}^*}{\partial (x_S^*)^2}$$

Multiplying the above equation by L/V^2 and simplifying yields

$$\frac{\partial v_{x,S}^*}{\partial t^*} + v_{x,S}^*\frac{\partial v_{x,S}^*}{\partial x_S^*} = -\frac{\partial p_S^*}{\partial x_S^*} + \frac{g_x L}{V^2} + \frac{\mu}{\rho L V}\left(\frac{\partial^2 v_{x,S}^*}{\partial (x_S^*)^2}\right)$$

Now consider the one-dimensional Navier–Stokes equation for the larger process, identified by the subscript L: it is

$$\frac{\partial v_{x,L}}{\partial t} + v_{x,L}\frac{\partial v_{x,L}}{\partial x_L} = -\frac{1}{\rho}\frac{\partial p_L}{\partial x_L} + g_x + \frac{\mu}{\rho}\frac{\partial^2 v_{x,L}}{\partial (x_L)^2}$$

We can convert this Navier–Stokes equation into a dimensionless equation just as before. Doing so give us

$$\frac{\partial v_{x,L}^*}{\partial t^*} + v_{x,L}^*\frac{\partial v_{x,L}^*}{\partial x_L^*} = -\frac{\partial p_L^*}{\partial x_L^*} + \frac{g_x L}{V^2} + \frac{\mu}{\rho L V}\left(\frac{\partial^2 v_{x,L}^*}{\partial (x_L^*)^2}\right)$$

Note both dimensionless equations have the same dimensionless groups, namely

$$\frac{g_x L}{V^2} \quad \text{and} \quad \frac{\mu}{\rho L V}$$

which are the inverse Froude number and the inverse Reynolds number. The Froude number is the ratio of the inertial forces to gravitational forces and the Reynolds number is the ratio of the inertial forces to viscous forces. Thus, if

$$\left(\frac{g_x L}{V^2}\right)_S = \left(\frac{g_x L}{V^2}\right)_L$$

and

$$\left(\frac{\mu}{\rho L V}\right)_S = \left(\frac{\mu}{\rho L V}\right)_L$$

then the two processes are mechanically equivalent.

However, the two processes must be geometrically similar for them to be mechanically similar. For each process, we defined

$$x_S^* = \frac{x_S}{L} \quad \text{and} \quad x_L^* = \frac{x_L}{L}$$

Thus

$$L = \frac{x_S}{x_S^*} \quad \text{and} \quad L = \frac{x_L}{x_L^*}$$

Equating the above equations, then rearranging give us

$$\frac{x_S}{x_S^*} = \frac{x_L}{x_L^*}$$

$$\frac{x_S}{x_L} = \frac{x_S^*}{x_L^*}$$

Therefore, the two processes are geometrically similar.

In summary, two processes are similar if their dimensionless geometric ratios are equal and if their dimensionless process parameters are equal. In other words, each process will generate a set of dimensionless Π parameters. When corresponding parameters are equal, then the comparator processes are similar. Symbolically

$$\Pi_1^{Geometric} = \Pi_2^{Geometric}$$
$$\Pi_1^{Static} = \Pi_2^{Static}$$
$$\Pi_1^{Kinematic} = \Pi_2^{Kinematic}$$
$$\Pi_1^{Dynamic} = \Pi_2^{Dynamic}$$
$$\Pi_1^{Thermal} = \Pi_2^{Thermal}$$
$$\Pi_1^{Chemical} = \Pi_2^{Chemical}$$

Thus, similarity depends upon Dimensional Analysis.

8.4 THEORY OF MODELS

The most general equation we can write for a prototype is

$$\Pi_1^P = f(\Pi_2^P, \Pi_3^P, \ldots, \Pi_n^P)$$

where the subscript numeral identifies a dimensionless parameter and superscript P indicates prototype. This equation applies to all mechanisms or processes that are comprised of the same dimensional variables. Thus, it applies to any model of the same mechanism or process, which means we can write a similar equation for that model

$$\Pi_1^M = f(\Pi_2^M, \Pi_3^M, \ldots, \Pi_n^M)$$

Dividing the prototype equation by the model equation gives us

$$\frac{\Pi_1^P}{\Pi_1^M} = \frac{f(\Pi_2^P, \Pi_3^P, \ldots, \Pi_n^P)}{f(\Pi_2^M, \Pi_3^M, \ldots, \Pi_n^M)}$$

Note that if $\Pi_2^P = \Pi_2^M$ and $\Pi_3^P = \Pi_3^M$ and so on, then

$$\frac{\Pi_1^P}{\Pi_1^M} = 1$$

Thus, $\Pi_1^P = \Pi_1^M$, which is the condition for predicting prototype behavior from model behavior. The conditions

$$\Pi_2^P = \Pi_2^M$$

$$\Pi_3^P = \Pi_3^M$$

$$\vdots$$

$$\Pi_n^P = \Pi_n^M$$

constitute the design specifications for the prototype from the model or the model from the prototype, depending whether we are upscaling or downscaling. If all these conditions are met, then we have a true model. If the above conditions hold for the controlling regime of the model and the prototype, then we have an adequate model. If most of the above conditions hold, then we have a distorted model that requires a correlation to relate Π_1^P and Π_1^M; in other words, we need an additional function such that

$$\Pi_1^P = f(\text{correlation})\Pi_1^M$$

If none of the above conditions holds true, then we have an analogous model.

We generally do not build true models in the chemical processing industry because the processes are so complex. A true model of a chemical process implies building a commercial-sized plant, which is far too costly and time consuming for an organization to do. Most models in the chemical processing industry are adequate or distorted models. Of these two types, adequate models are the better since they model the controlling regime of the process. Distorted models are more difficult to use because we have to determine the correlation between the distorted model and the prototype. Developing that correlation takes time and costs money—two commodities in short supply in our global economy.

8.5 SUMMARY

This chapter demonstrated the dependence of scaling and model theory upon Dimensional Analysis. It also discussed the types of models available to chemical engineers.

REFERENCES

1. V. Skoglund, *Similitude: Theory and Applications*, International Textbook Company, Scranton, PA, 1967, pp. 74–75.

2. R. Johnstone, M. Thring, *Pilot Plants, Models, and Scale-Up Methods in Chemical Engineering*, McGraw-Hill, New York, NY, 1957, Chapter 3.

3. R. Granger, *Fluid Mechanics*, Dover Publications, New York, NY, 1995, p. 204.

CHAPTER 9

An Assessment of Dimensional Analysis

9.1 SCOPE OF DIMENSIONAL ANALYSIS[1]

The power of Dimensional Analysis lies in its ability to provide a functional relationship describing a complex mechanism or process. In order to do that, we must understand the physics of the mechanism and the chemistry of the process; otherwise, we cannot successfully identify the variables involved in the mechanism or process. In such situations, Dimensional Analysis cannot be done.

If we have a partial understanding of the physics and chemistry involved in a given mechanism or process, then we can identify a group of variables comprising the mechanism or process. We can use such a group of variables for Dimensional Analysis; however, the result will be incorrect if we missed an important variable or misleading if we used incomplete information.

If we know all the relevant variables impacting a given mechanism or process, then Dimensional Analysis produces a valid functional solution that provides valuable physical insight. However, experiments must be done to determine the equation of the solution function. These experiments may identify controlling regimes in the mechanism or process. Such knowledge provides us with information for simplifying the conservation laws and constitutive equations describing the mechanism or process, thereby allowing an analytic solution to be developed.

If an analytic solution exists for a given mechanism or process, then we need not use Dimensional Analysis to generate a functional solution.

9.2 USES FOR DIMENSIONAL ANALYSIS[2]

We use the concepts of Dimensional Analysis in a variety of ways. We use the concept of dimensional homogeneity as a check on our algebraic manipulations. If a physical equation involves apples and we finish with "orpels," then we know we have made an incorrect algebraic manipulation in our study. We also use Dimensional Analysis to develop self-consistent systems of units and to establish the conversion factors between those systems of units.

As mentioned several times previously, aeronautical, civil, and mechanical engineers have used Dimensional Analysis quite success-fully to obtain functional solutions to problems too complex for deri-vation of an analytic solution. The problems these engineering disciplines have solved using Dimensional Analysis involve the smallest possible set of fundamental dimensions and a small number of vari-ables. Some heat transfer problems have been resolved using Dimensional Analysis. Fewer such solutions exist because they involve a more complex set of fundamental dimensions and a larger number of variables. Chemical processes are so complex that few attempts have been made to use Dimensional Analysis to analyze them. This situa-tion will change in light of free-for-use matrix calculators on the Internet.

Once we have a functional solution developed from Dimensional Analysis, we can use it to gain physical insight about the problem. For example, using a functional solution, we can generate experimental data identifying the controlling regimes of the mechanism or process.

Dimensional Analysis provides us with a tool for designing experi-mental programs. By grouping dimensional variables into dimension-less parameters, we can reduce the number of independent variables, thereby reducing the number of experiments required to define a mech-anism or process. Using dimensionless parameters also allows us to succinctly present the results of a given experimental program.

Lastly, Dimensional Analysis provides a firm, logical foundation for the theory of models. Thus, Dimensional Analysis provides us with a method for upscaling or downscaling complex mechanisms or processes.

9.3 LIMITATIONS OF DIMENSIONAL ANALYSIS[2]

Dimensional Analysis has significant limitations. Those limitations are

- it investigates physical quantities; thus, it uses intrinsic properties and not physical magnitudes;

- it does not produce numerical results;
- it does not provide analytic solutions to the conservation laws and constitutive equations governing a mechanism or process;
- it does not indicate whether important physical quantities or variables are missing from a functional solution;
- it does not indicate whether the correct set of physical quantities or variables are used in a particular functional solution;
- it does not produce unique solutions;
- it produces solutions requiring experimental confirmation.

9.4 SUMMARY

This chapter discussed the scope, uses, and limitations of Dimensional Analysis.

REFERENCES

1. M. Zlokarnik, *Scale-Up in Chemical Engineering*, Second Edition, Wiley-VCH Verlag GmbH & Co. KGaA, Weinheim, Germany, 2006, pp. 52–53.

2. R. Pankhurst, *Dimensional Analysis and Scale Factors*, Chapman & Hall, London, UK, 1964, Chapter 8.